Earth Science

Reviewing the Essentials

Thomas McGuire
Former Science Coordinator
Briarcliff High School
Briarcliff Manor, New York

AMSCO

AMSCO SCHOOL PUBLICATIONS, INC.
315 Hudson Street, New York, N.Y. 10013

Illustration and Photo Credits
Figure 16-1 Adapted from *Pennsylvania and the Ice Age*.
Figure 26-3 Courtesy of Hubble Heritage Team (STScI/AURA/NASA).
Figure 28-6 Courtesy of NASA.

Cover and text design by Merrill Haber
Artwork by Hadel Studios
Composition by Northeastern Graphic Services, Inc.

The publisher wishes to acknowledge the contributions of **Neal Grasso**,
Geologist, Boston, Massachusetts, in the preparation of this book.

When ordering this book, please specify:
either **R 704 P** *or* EARTH SCIENCE: REVIEWING THE ESSENTIALS

ISBN-13 978-0-87720-182-3

Please visit our Web site at: www.amscopub.com

Printed in the United States of America

30   12

PREFACE

Recent initiatives in science education, including standards from the American Association for the Advancement of Science and the American Geological Institute, as well as other respected organizations, have guided the preparation of this book. Many school districts are aligning their curricula with national standards, while states are adopting assessments to test schools' progress in meeting these standards.

The following principles are common to most of these standards:

A. Science is a dynamic process. The factual content of science is a temporary product of observations and experimentation. As we propose new ideas and develop better technology, we expect the content of science to change.

B. Science literacy is an important part of the education of all citizens. While few students will become scientists, most will become participants in our political and economic systems. To make the best personal and civic choices, they will need to be familiar with scientific issues.

C. Planet Earth is our most immediate science laboratory. It is what we investigate, where we investigate, and it guides us in how to investigate our surroundings.

Earth Science: Reviewing the Essentials is written for users of diverse abilities and backgrounds. The material is appropriate for students in an introductory course in earth science. Some of the content may require a familiarity with a few basic principles of algebra and geometry, but this quantitative material can be utilized as extensions or optional excursions.

This book should be accompanied by a laboratory/student activity program. Although every effort has been made to provide clear and concise explanations, many concepts covered should be mastered in a more active mode. After all, at its best, learning science is doing science.

Many teachers will select a full-featured textbook as the primary source for their course of study. Others may require students to read selected references. Whatever your teacher's choice, *Earth Science: Reviewing the Essentials* is designed to convey a broad range of basic content and intellectual skills.

The outstanding features of this book are as follows:

1. Presentation: The content of this review text has been organized into 28 short chapters. The chapters are written in a simple narrative style that makes the content more easily understood. Important words to remember are **boldfaced** to make them stand out. These words are defined in the text where they appear for the first time. It might help you to remember these words if you copied them into your notebook along with their definitions. Then you can use this list when you review for an exam. The index allows you to locate key words and topics quickly.

2. Illustration program: When appropriate, line illustrations are integrated in the text. These illustrations present a different way for you to learn. They are drawn by artists who are skilled in presenting scientific material in an accurate and interesting manner. You may want to copy some of these illustrations in your notebook. It really does not matter if you are a skilled artist —these drawings are meant to help you learn and organize the earth science information you will be taught this year. Charts, graphs, and tables are also included throughout this text. These graphics present another way to process information that many students find useful. Information in the charts and tables is usually presented in a brief way. The "key words" included help students organize and synthesize the data presented. You may wish to copy the charts and tables into your notebook. Your teacher may wish to write the charts and tables on the chalkboard. The illustrations in this review text are an integral part of the program—you should use them as additional study aids.

3. Questions: The questions presented at the end of each chapter help you master the material presented. If you answer a question incorrectly, you should reread the section of the text that covered that material. The questions in this review text are not meant to mislead you. They are written to test your understanding of the material, and to help you recognize gaps in your knowledge so that you can review that topic. A variety of different question types are included, such as multiple-choice, matching, fill-in, essay, and portfolio questions. These different types of questions will help you review as you prepare for exams.

Earth Science: Reviewing the Essentials will prove to be an effective aid for students in mastering the material presented in a one-year earth science course. Earth surrounds you, and includes you. This brief review text will help you explore our Earth—a planet on which you play an important part.

CONTENTS

Chapter 1
The Science of Planet Earth

A WONDERFUL INHERITANCE has been left to you. It is in the form of a very special planet called Earth. This planet is covered with an assortment of natural wonders—mountains, canyons, rivers, oceans, deserts, forests, and much more. However, Earth did not come with a guidebook or set of directions. Instead, we must all discover the laws of nature for ourselves. In this process of discovery, we will make mistakes. But our ultimate goal should be to preserve the planet during our stay on it. We can begin to investigate the features of Earth through the pursuit of science.

THE NATURE OF SCIENCE

What do you think of when you hear the word *science*? From your experiences in school, you may think of science as a complete body of knowledge. In fact, the Latin origin of the word *science* (*scire*) means "to know." But this definition is only partly correct because facts and other information are not the true heart of this discipline. **Science** is an organized approach to gathering, analyzing, verifying, and utilizing information about the world. Rather than an unchanging body of knowledge, science is an unending process.

The process of science begins with a simple observation followed by a question. Perhaps you have wondered why the sky is blue, why we have four seasons, or why the moon revolves around Earth. Maybe you have seen photographs of the damage

**Figure 1-1. Science is the key to
understanding our world.**

caused by a volcano, hurricane, or earthquake and wondered
how it all happened. When your observations lead you to ask
such questions, you are thinking like a scientist. Scientists observe the world around them and ask questions.

What do you do when you have a question? Naturally, you
look for an answer. Scientists also try to find answers to their
questions. They do so by gathering information and formulating
a **hypothesis**, or suggested answer to a question. A hypothesis is
usually tested through experimentation, in which additional observations are made and measurements are recorded. The results of the experiment are analyzed to determine whether they
support or disprove the hypothesis. The scientist then draws a
conclusion based on the research.

You may think that once a scientist draws a conclusion, the
process of science comes to an end. But this is not the case. A
fundamental characteristic of good science is the ability to re-
produce the results. In other words, another person doing the
same experiment under similar conditions should observe similar results. If the results cannot be repeated, the results may
have been due to inaccurate observations, poor experimental
procedures, or random chance. There are many documented
cases in which scientists were too quick to release news of their

discoveries. When their peers repeated the experiment, they were unable to duplicate the results.

Once an experimental result has been reproduced many times, it may be set forth as a theory. A **theory** is a logical explanation based on data gathered and verified by many scientists. If at some point new data or experiments do not agree with the theory, the theory must be changed or discarded.

The fact that a theory can be changed is one of the beauties of science. The process of science enables us to continually assess our understanding and change our ideas as we improve our knowledge of the universe. Our beliefs about nature must be as flexible as nature itself. Ideas about nature cannot be refined to the point that they become unchanging principles. Science simply brings us to our current understanding of nature.

WHAT IS EARTH SCIENCE?

The study of science can be classified into three branches: life science (such as biology), physical science (such as physics and chemistry), and earth science. This book is dedicated to the study of earth science. **Earth science** is the systematic study of solid Earth, the water on it, and the air around it, as well as Earth's place in the universe.

Earth science can be further divided into several branches. **Geology** is the study of solid Earth, including Earth's surface and interior. A geologist studies the origin, history, and structure of Earth and the processes that shape its surface. Some geologists identify minerals. **Meteorology** is the study of Earth's atmosphere and its changing conditions. By studying the atmosphere, meteorologists can learn about weather and climate. **Oceanography** is the study of the characteristics and dynamics of Earth's oceans, which cover about 70 percent of the planet. **Astronomy** is the study of the motions of planet Earth and objects outside Earth. In other words, astronomy is the study of the universe. One more branch of earth science that is closely related to the other branches is ecology. **Ecology** is the study and protection of the environment.

Although it is helpful to divide the study of science and earth science into branches, no scientist works in isolation. The

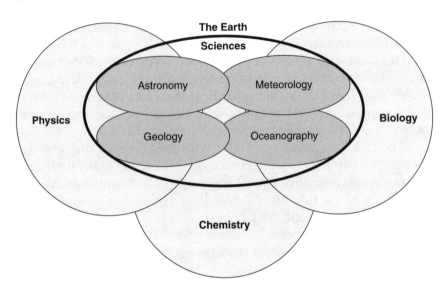

Figure 1-2. Although each branch of science is different, they are all related.

various sciences constantly overlap. Most of the great discoveries in history have occurred as a result of scientists from many branches working together.

WHY STUDY EARTH SCIENCE?

No other species has the ability to use and modify Earth as humans do. As a result, Earth is being subjected to ever-increasing stresses from growing populations and the rapid expansion of technology. Limited resources are being used or polluted at alarming rates. You don't have to be a professional scientist to take an active role in determining the fate of the planet. As citizens of a democracy, and active participants in a world economy, we have the opportunity and the responsibility to protect and preserve the environment on planet Earth.

If you think you cannot make a difference, you are wrong. Prior to the 1970s, for example, the use of leaded gasoline released poisonous substances into the environment. Pressure from the public resulted in the elimination of the use of leaded gasoline and the establishment of stricter air quality laws over the past several decades. Similarly, recycling of materials rather than burning or burying them is the result broad public pressure. These successes and many others are, in large measure, the result of education. In order to make intelligent decisions regarding the planet, both in government and in business, people need to understand Earth systems. Keep this in mind as you study earth science.

SCIENTIFIC OBSERVATIONS

We get information about Earth through our senses. The senses are our gateway to the world, and it is not possible to pursue science without making extensive use of the senses. A piece of information obtained directly through our senses is called an **observation**.

Some observations are **qualitative**. These observations include information that is relative but not specific. For example, if you were told that a person is very tall, you might picture a person who stands above her friends, but you would not know her actual height. Similarly, when you note that the color of a flower is red, you are making a qualitative observation. Qualitative observations do not involve measurements. Scientific instruments that can be used to make qualitative observations include microscopes, telescopes, and binoculars.

When you count objects or make a measurement, you are making a **quantitative** observation. The word *quantitative* is based on the word *quantity*, which means "how many." If you know that a person is 1.7 meters tall, you know her exact height. Scientific instruments that can be used to make quantitative observations include metersticks, balances, and stopwatches.

SYSTEMS OF MEASUREMENT

All quantitative values are expressed with numbers and units. Units are accepted values for the quantity being expressed. In the United States, we use a variety of units. For example, distance can be expressed in such units as inches, feet, yards, fathoms, furlongs, statute miles, and nautical miles. These particular

Table 1-1.   SI Units of Measurement

Quantity Measured	Unit	Symbol
Mass	gram	g
Length	meter	m
Volume	liter	L
Temperature	degrees Celsius	°C
Time	second	s

units are related by multiples of 12, 3, 2, 110, 8, and 1.150779...
respectively. That is, 12 inches in 1 foot, 3 feet in 1 yard, 2 yards
in a fathom, 110 fathoms in a furlong, 8 furlongs in a statute
mile, and 1.150779... statute miles in a nautical mile.

This mix of units has been replaced in most countries by the
International System of Measurement (SI). SI is a decimal
system of measurement. Units within SI are related in multiples
of 10. The basic unit of length is the meter (m). A meter is a little
longer than a yard. To measure a length smaller than a meter,
you can use centimeters (cm). There are 100 cm in a meter. For
even smaller lengths, the millimeter (mm) is used. There are 10
millimeters in a centimeter and 1000 millimeters a meter. The
kilometer (km) is used for long distances. There are 1000 meters
in a kilometer. Table 1-1 lists some of the most commonly used
SI units of measure. Table 1-2 lists the prefixes used to denote
larger and smaller units.

Table 1-2.   Commonly Used Prefixes in SI

Prefix	Meaning
Mega-	1,000,000 times
Kilo-	1000 times
Hecto-	100 times
Deka-	10 times
Deci-	1/10th
Centi-	1/100th
Milli-	1/1000th
Micro-	1/1,000,000th
Nano-	1/1,000,000,000th

QUESTIONS earth is ≈100,000 Kilometer
Multiple Choice Atmostsphere in circumference

1. The first thing you need to do as you begin a scientific investigation is to   *a.* obtain measuring instruments *b.* decide who will perform each part of an experiment *c.* publish your theory   *d.* make an observation and ask a question.

2. A preliminary answer to a question is a   *a.* hypothesis *b.* theory   *c.* conclusion   *d.* fact.

3. The branch of earth science in which you might study the changing phases of the moon is   *a.* meteorology *b.* geology   *c.* astronomy   *d.* oceanography.

4. An earth scientist might do all of the following except *a.* find ways to observe and photograph atoms   *b.* discover how carbon dioxide affects our climate   *c.* map the bottoms of the world's oceans   *d.* measure how fast the sun appears to move through the sky.

5. Which event might be studied by a meteorologist?   *a.* a strong earthquake   *b.* a nighttime shower of "shooting stars"   *c.* a large forest fire   *d.* an intense hurricane

6. Two teams of scientists conduct the same experiment. How would you expect their results to compare?   *a.* They would have exactly the same data.   *b.* Their data should lead to similar conclusions.   *c.* Their conclusions should be very different.   *d.* It would not be possible to compare their findings.

7. An example of a qualitative statement is   *a.* the fossil is two million years old   *b.* there are nine planets in our solar system   *c.* light from distant stars takes many years to reach Earth   *d.* the river is 153 kilometers long.

8. Which is a quantitative statement?   *a.* Earth is not as old as the universe.   *b.* Stars give off light because they are sources of energy.   *c.* The density of water is 1 gram per cubic centimeter.   *d.* Precision measurements may require the use of special instruments.

9. A metric unit of length is the   *a.* gram   *b.* meter *c.* degree Celsius   *d.* liter.

Matching

_____b_____ 10. Geology

_____c_____ 11. Oceanography

_____d_____ 12. Meteorology

_____a_____ 13. Astronomy

a. Branch of earth science that might involve locating the North Star, Polaris

b. Branch of earth science that might involve studying changes in Earth's rocks over time

c. Branch of earth science that might involve measuring the chemical content of salt water

d. Branch of earth science that might involve predicting next week's weather

Fill In

14. Discovery in science is valid only if other scientists can _explain_ the results.

15. The most fundamental activity of science is _experiment_

16. Astronomy, geology, meteorology, and oceanography are grouped together as _Earth science_

17. To find Earth's buried natural resources, look to the science of _____.

18. _____ is the branch of earth science that deals primarily with environmental issues.

19. A piece of information that we get through our senses is known as a(an) _____.

20. _____ is a system of measurement based on the number 10.

Free Response

21. A friend says that *science* is in a book. Agree or disagree with this statement and explain your choice.

22. Why is it an important feature of science that all our understanding of the world be flexible?

23. Explain why it is important to study science even if you do not expect to become a professional scientist.
24. Why is the ability to duplicate the research of another scientist an important characteristic of science research?
25. Compare and contrast qualitative and quantitative observations.

Portfolio

I. An example of a discovery that was not researched according to proper scientific procedures involved cold fusion. Conduct library research about the event and write a report explaining why the findings should not have been published so quickly.
II. Prepare a listing of professional titles or occupations that fall within the realm of earth science.
III. Construct a table with columns for the pros and cons of SI units. Then fill in the table with reasons why the United States should, or should not, abandon our customary units of measure.

Chapter 2
The Dimensions of Planet Earth

Eᴀʀᴛʜ ɪꜱ ᴀɴ ᴀᴍᴀᴢɪɴɢ ᴘʟᴀɴᴇᴛ characterized by many different features. Where do you begin a study of Earth? There is no right answer, but it is helpful to consider Earth as a whole before identifying specific parts to study. In this chapter, you will learn about Earth's shape, size, and subdivisions.

THE SHAPE OF EARTH

You know that Earth's shape is a ball, or sphere. In fact, you have probably seen photos of Earth taken from space that show Earth as a sphere. Although this fact has been accepted for several hundred years now, it was not always obvious. After all, when you look out at the horizon, Earth looks flat. So for a long time, many people believed Earth was flat.

Despite the general misunderstanding about Earth's shape, there were some people who knew both the shape and size of planet Earth more than 2000 years ago—long before Columbus set sail. They deduced Earth's true shape from observations of natural phenomena. One piece of evidence came from observing sunlight. Even after the sun set over a flat horizon, sunshine was observed on nearby mountains and clouds. If Earth were flat, this would be impossible because the sun could be in one of two places—either above or below Earth.

Another piece of evidence came from eclipses. During an eclipse of the moon, Earth moves between the sun and the moon.

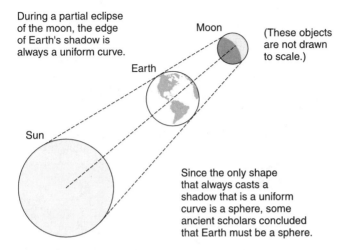

During a partial eclipse of the moon, the edge of Earth's shadow is always a uniform curve.

Moon

(These objects are not drawn to scale.)

Earth

Sun

Since the only shape that always casts a shadow that is a uniform curve is a sphere, some ancient scholars concluded that Earth must be a sphere.

Figure 2-1. An eclipse of the moon provides evidence for Earth's shape.

When this happens, Earth casts a shadow on the moon. Observers noticed that the shadow traced a circle on the moon as only a sphere would do. (You will learn more about eclipses in Chapter 28.)

An additional piece of evidence came from observations of stars made by sailors. Early sailors found that the farther north they traveled, the higher in the sky the North Star, Polaris, rose. As they traveled south, Polaris seemed to move lower in the sky. The gradual change in position of the star could mean only that the ship was moving on the surface of a sphere.

Finally, people who watched ships leave or arrive in port discovered one more piece of evidence about Earth's shape. Observers noticed that the mast, or vertical pole, was the last part of the ship to disappear as a ship left port. It was also the first part to be seen as the ship approached. This indicates that Earth's surface is curved. If Earth were flat, you would see all parts of the ship at the same time.

AN IMPERFECT SPHERE

In reality, Earth is not a perfect sphere. Notice that Earth is slightly flattened at the poles and bulges at the equator. The

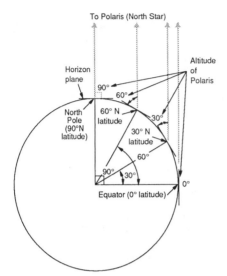

Figure 2-2. As an observer moves northward on Earth's surface, Polaris moves higher above the horizon.

equator is an imaginary line around Earth located halfway between the poles. Earth bulges because it rotates, or spins, on its axis. Earth's **axis** is an imaginary straight line through the planet between the North Pole and the South Pole. The bulge of Earth's equator makes the diameter at the equator about 40 kilometers greater than the diameter through the poles. This shape is called an oblate spheroid. For most purposes, it is acceptable

Figure 2-3. Due to the curve of Earth's surface, the mast of a ship is the last part to disappear as the ship sails away.

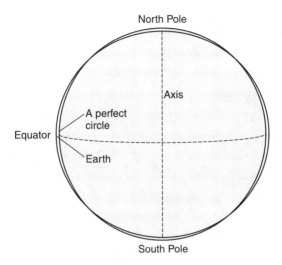

Figure 2-4. Earth bulges very slightly at the equator as the planet spins on its axis.

to approximate the shape of Earth as a sphere. The reason is that the bulge is tiny compared to the overall size of Earth—less than 1 percent of Earth's diameter. So the planet looks like a perfect sphere from space.

GRAVITY

How do scientists know that Earth is not a perfect sphere if the bulge is so small? One way is by studying gravity. **Gravity** is the force of attraction that every object in the universe exerts on every other object. The force of gravity between two objects depends on mass and distance. Mass is the quantity of matter in an object. Gravity is directly proportional to mass and inversely proportional to distance. That means that the greater the mass of two objects, the greater the force of gravity between them. The greater the distance between two objects, the smaller the force of gravity between them.

You are measuring the force of gravity between your body and Earth when you weigh yourself. **Weight** is equal to gravity (g) times mass (m). At any given time, your mass is constant, but your weight can change depending on the distance between you

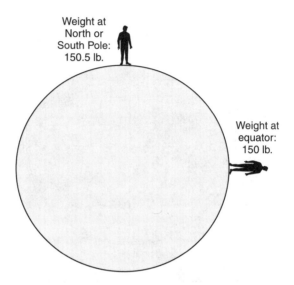

**Figure 2-5. The weight of a person or object
is slightly greater at the poles than at the
equator.**

and the center of Earth. Careful measurements have shown that
the weight of an object is slightly greater at the poles than at the
equator. A greater weight for the same mass means a greater
force of gravity. Since gravity depends on mass and distance, the
greater force of gravity on the same mass means that the distance must be less. The distance from Earth's surface to its core
decreases as a mass moves from the equator to the poles.

THE SIZE OF EARTH

Measuring something as large as Earth is a difficult task.
You might think that you need advanced technology, but this is
not necessarily the case. The distance around Earth, its circumference, was first calculated by the Greek scholar Eratosthenes
(er uh TOS thuh neez) more than 2000 years ago. Eratosthenes
used geometry, which is the branch of mathematics that involves
shapes and angles. The word *geometry* comes from the Greek
roots *geo* meaning "Earth" and *metron* meaning "measurement."

Eratosthenes noted that the sun shone to the bottom of a
deep well at noon once a year—on the first day of summer. The

well was located at Syene, a city along the Nile River in southern Egypt. For the sun to shine on the bottom of the well meant that the sun was directly overhead. At the same exact time in Alexandria, a city thought to be directly north, the sun was about 7.2° away from the vertical. Since a circle is divided into 360°, 7.2° is one-fiftieth of a circle. By using the distance from Alexandria to Syene, Eratosthenes set up a proportion that enabled him to calculate the circumference of Earth.

During Eratosthenes' time, distance was measured in units called stadia. (The unit stadium was a standard length used for footraces. *Stadia* is the plural of *stadium*.) The distance between these two cities was estimated to be 5000 stadia.

$$\frac{7.2°}{360°} = \frac{5000 \text{ stadia}}{\text{Earth's circumference}}$$

$$7.2 \times \text{Earth's circumference} = 1,800,000 \text{ stadia}$$

$$\text{Earth's circumference} = 250,000 \text{ stadia}$$

Although modern scientists do not know the exact length of 1 stadium, it does appear that Eratosthenes' figure was remark-

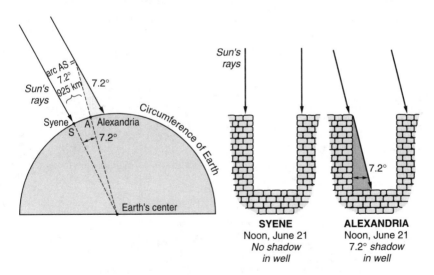

Figure 2-6. Eratosthenes calculated Earth's circumference based on observations of the sun's rays in two different cities.

Table 2-1. Earth's Measurements

	Diameter	*Circumference*
At poles	12,714 km	40,007 km
At equator	12,756 km	40,074 km

ably close to the true distance, now known to be about 40,000 kilometers. The method used by Eratosthenes is still used today. However, the data are gathered more precisely.

THE SUBDIVISIONS OF EARTH

It is convenient to divide Earth into subdivisions based on its three main features: land, water, and air. The land, or solid part of Earth, is known as the **lithosphere**. The lithosphere extends from the solid surface of Earth to a depth of about 100 kilometers. Although the properties of Earth change below the lithosphere, for now we will consider Earth as if it were a solid ball made up of a uniform material.

About 70 percent of the lithosphere is covered by water, known as the **hydrosphere**. The hydrosphere includes oceans, rivers and streams, ponds and lakes, and other bodies of water.

The **atmosphere** is the shell of gases that surrounds Earth. Although it extends more than 100 kilometers into space, most

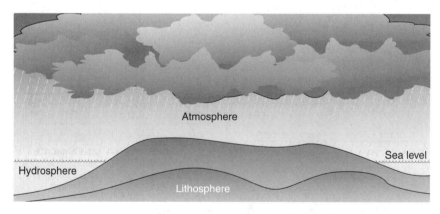

Figure 2-7. Major subdivisions of Earth.

of it is contained within the lower 10 kilometers. This is because the air is compressed by the weight of air above it, making the lowest layer of the atmosphere much denser than the layers above. The atmosphere has no top boundary; it just gets thinner and thinner as it extends into space.

The atmosphere is further divided into layers based on changes in temperature. The lowest layer is the **troposphere**, which becomes cooler as you rise toward the top. The gases of the troposphere are essential to life on Earth. In addition, most of Earth's weather occurs in the troposphere. The top boundary of the troposphere, at which the temperature decrease stops, is known as the **tropopause**. The next layer, the **stratosphere**, reaches from the tropopause to a height of about 50 kilometers from Earth. The stratosphere has strong, steady winds, such as the jet stream. The lower end of the stratosphere is as cold as the

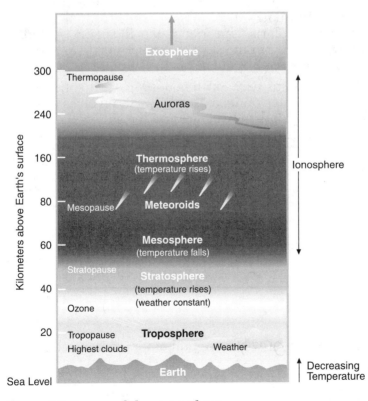

Figure 2-8. Layers of the atmosphere.

top of the tropopause, but it warms up as you rise to its top, the **stratopause**. The next two layers of the atmosphere are the **mesosphere**, in which temperature decreases with height, and the **thermosphere**, in which temperature increases with height.

There is one more subdivision on Earth, called the biosphere. All the living things of Earth are part of the **biosphere**. The biosphere is unique in that it exists entirely within the other spheres—lithosphere, hydrosphere, and atmosphere. The biosphere and all three of the other parts of Earth form one interdependent and dynamic system.

QUESTIONS

Multiple Choice

1. How do observations of Polaris support the idea that Earth is not flat? *a.* Polaris is visible from all places on Earth. *b.* Polaris appears circular in shape. *c.* Polaris appears higher in the sky the farther north you go. *d.* Polaris is the brightest star in the night sky.

2. Which explains why Earth is not a perfect sphere? *a.* Earth rotates on its axis. *b.* Earth revolves around the sun. *c.* Earth is moving around the galaxy. *d.* The universe is expanding.

3. Because the moon has less mass than Earth, your weight on the moon would be *a.* greater than on Earth *b.* less than on Earth *c.* the same as on Earth *d.* impossible to predict.

4. The state of matter found in Earth's hydrosphere is *a.* solid *b.* liquid *c.* gas *d.* plasma.

5. The oxygen you breathe is found in the *a.* lithosphere *b.* hydrosphere *c.* troposphere *d.* stratosphere.

Fill In

6. The force of gravity is the pull among all objects in the universe.

7. Weight changes with mass and weight of distance.

8. The equator is the solid part of Earth. lithosphere

9. The *hydrosphere* covers nearly three-quarters of solid Earth.
10. As you travel upward from Earth's surface, the density of air *thins*.
11. The living parts of planet Earth, the *biosphere*, exist within the three other major spheres.
12. The atmosphere is divided into sections according to changes in *gases./temperature*

Free Response

13. How do observations of ships sailing out to sea support the idea of a round Earth?
14. If some of Earth's mountains rise 10 kilometers high, why does Earth look so round and smooth from space?
15. Why do objects weigh slightly less at the equator than they do at the poles?
16. How did Eratosthenes calculate the circumference of Earth?
17. Why is the lowest layer of the atmosphere the most dense?

Portfolio

Contact students at a school located directly north or south in another state to find the size of Earth by measuring the angle of the sun in the sky at the same time at both schools and the distance between the two schools.

Chapter 3
Maps and Navigation

HAVE YOU EVER USED a map to get from one place to another? A map is a drawing of Earth or part of Earth. There are many different types of maps, and they come in all shapes and sizes. The information in a map depends on the purpose for which the map was designed. Some maps show the positions and shapes of land. Other maps show such information as political boundaries, highways, geologic features, weather systems, or population centers. In this chapter you will learn how maps are made and used. You will also find out how to navigate without a map.

MAP PROJECTIONS

The most accurate maps of Earth are globes because they have the same shape as Earth. A globe, however, is not very convenient to carry or store. For practical purposes, flat maps are most often used instead of globes. Showing a spherical surface accurately on a flat map is a difficult task. Imagine trying to flatten the outside of a tennis ball. To make it flat you would have to cut the ball. This would change, or distort, the shape of the ball. In a similar way, flat maps distort shapes, distances, or directions when they represent Earth. Mapmakers have developed ways for showing the curved surface of Earth on a flat map. Such maps are known as **map projections**. Some map projections show true shapes but distort distances and directions. Others show true distances and directions but distort shapes.

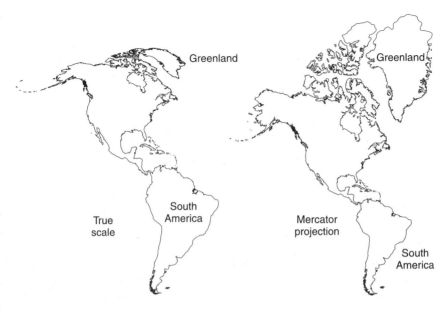

Figure 3-1. Although a Mercator map shows the correct directions, it distorts the size of features near the poles. Notice how Greenland, actually much smaller than South America, looks larger on a Mercator map.

One type of map projection is the **Mercator projection**, which is convenient because it shows all of Earth in one continuous map. One major problem with a Mercator projection is that regions near the poles are enlarged tremendously. With this common map projection, Greenland looks as large as South America when it actually is much smaller.

Each map projection has advantages and disadvantages. The best map for one use may not be the best for another use. The more familiar you are with the various kinds of maps the better your selection of a map for a particular task.

TERRESTRIAL COORDINATES

No matter what type of map you use, you must be able to locate specific positions on Earth. The coordinate system makes this possible. In this system, a grid of lines is drawn on a map of

Earth. Every position on Earth can be described in reference to these lines by giving a pair of numbers, or coordinates, called latitude and longitude. It is much like using the x- and y-coordinates on a graph. All maps show the same latitude and longitude for a particular point.

Latitude is based on the reference line known as the equator. The angular distance north or south of the equator is called **latitude**. This angle is always measured to the center of Earth. Lines of latitude, or **parallels**, are drawn parallel to the equator. Latitude measurements extend from 0° at the equator to a maximum of 90° at the North and South poles.

Latitude lines are spread out evenly on Earth's surface. The distance between two lines of latitude is about 112 kilometers. This is true because a circle can be divided in 360°. One degree of latitude is 1/360 of Earth's circumference at the poles. Since the circumference is about 40,000 kilometers, a degree of latitude is about 112 kilometers. Each degree of latitude is further divided into 60 minutes, which are about 1.85 kilometers each. Each minute is divided into 60 seconds.

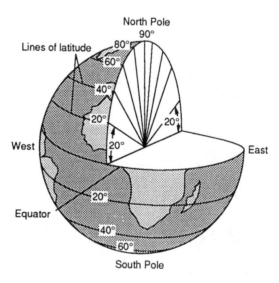

Figure 3-2. Latitude is the angle north or south of the equator.

Latitude describes distances north and south on Earth's surface. The second coordinate, longitude, describes distances east and west. Unlike latitude, which uses the equator for a reference line, there is no obvious reference line to use for longitude. For this reason meridians have been established. A **meridian** is an imaginary line that circles Earth from the North Pole to the South Pole. The meridian running through Greenwich, England, is called the **prime meridian**. The angular distance east or west of the prime meridian is called **longitude**. The longitude of the prime meridian is set at 0°. Like latitude, longitude is an angle measured from the center of Earth. You can move east or west of the prime meridian up to 180°. The half of Earth west of the prime meridian has west longitude; the other half has east longitude.

Like latitude, longitudinal degrees can be further divided into minutes and seconds. Unlike parallels, however, meridians are not spaced evenly. They are closer together the farther they are from the equator. There is no set distance for each degree. Instead, they become smaller and smaller.

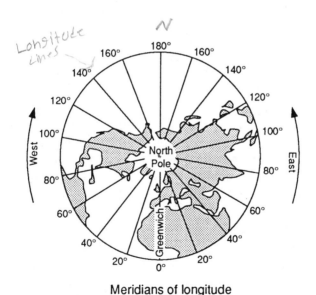

Meridians of longitude

Figure 3-3. Longitude is the angle east or west of the prime meridian.

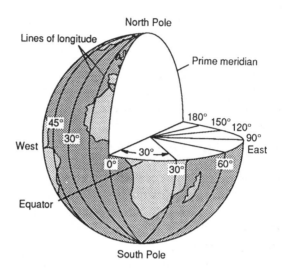

North Pole

Lines of longitude

Prime meridian

180° 150° 120°

45° 90°

West 30° East

30°

0° 30° 60°

Equator

South Pole

Figure 3-4. Lines of longitude are closest near the poles and farthest apart at the equator.

MAP SCALES

Not all maps are the same size. A map of the United States, for example, might be as large as a wall or as small as a notebook. In addition, not all maps represent the same portion of Earth. One map might represent a country, and another might represent one city within the country. So how can you tell distance on a map? A map **scale** relates the distance on the map to the distance on Earth. The closer the size of the map to the land it represents, the larger the map's scale.

Map scales are usually expressed in one of three ways. The first way is as a general statement. For example, "1 centimeter represents 50 kilometers." The second way is to print a numerical statement such as the fraction 1 cm/50 km or the ratio 1 cm: 50 km. This is the same scale as in the previous sentence but stated in a different way. In using this kind of scale, you can measure the distance between two locations in centimeters and divide by 50 to change to the number of kilometers that it represents on land.

The third way to show map scale is by a line divided into equal lengths. The line is marked in units of length such as kilo-

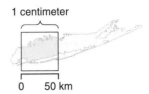

1 centimeter

0   50 km

Figure 3-5. The scale of this map of Long Island is 1 centimeter to 50 kilometers.

meters or miles. To determine the real distance between any two points on the map, place a sheet of paper on the map with the edge touching the two points in question. Make two marks on the paper to show the map distance between these points. Then move the paper next to the scale line to determine the actual distance.

TOPOGRAPHIC MAPS

When you look at the map in Figure 3-1, you see land and water, but you cannot identify mountains and valleys. In other words, you cannot use the map to learn about **topography**, which is the shape of Earth's surface. A map that shows the three-dimensional shape of Earth on a flat surface is called a **topographic map**. The highs and lows of a land surface are also known as the **relief** of the land. Topographic maps use contour lines to show relief. **Contour lines** connect places that have the same elevation. Elevation is the height above sea level. You can think of a contour line as a step up or down to a new level. Contour lines never cross.

The numbers on the contour lines are measured elevations. Some topographic maps use different colors for different elevations. The difference in elevation between two consecutive contour lines is called the **contour interval**. For example, a map with a contour interval of 10 meters will show contour lines at elevations of 0 meters, 10 meters, 20 meters, and so on. On any map, the size of the contour interval should be the same everywhere. The contour interval for a particular map depends on the relief of the land.

Look at the landscape shown in Figure 3-6. You may find the map a little confusing at first. It will become easier to read as you learn more about topographic maps. With a little prac-

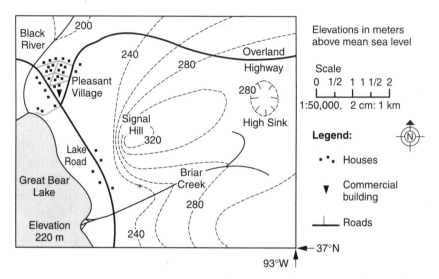

Figure 3-6. A topographic map shows Earth's three-dimensional shape on a flat surface.

tice, you will be able to recognize landscapes, such as hills and valleys, from the shape of the contour lines. A steep slope, for example, is represented by closely spaced contour lines. The lines are close together because the elevation of a steep slope changes greatly over a short distance. A gentler slope is represented by contour lines that are spaced far apart because the elevation changes only slightly over a short distance. Contour lines usually form a V-shape in a stream valley. They also form small, closed loops around hilltops or depressions. You can tell whether the loops represent a hilltop or a depression by looking at the elevation numbers. If the numbers increase toward the center of the closed loop, the feature is a hilltop. If the numbers decrease, the feature is a depression. Most topographic maps also show natural features, such as streams and lakes, as well as many features made by humans, such as roads, trails, and even some buildings. A key must be provided to describe the symbols used in the map.

The United States Geological Survey publishes detailed topographic maps for all parts of the United States. You may want to use your local topographic map for locating places or planning travel. These maps are available from the United States government as well as from many local hardware and sports shops.

A topographic map is a specific type of isoline map. An **isoline map** shows an area in which the same type of measurement has been made in several locations. On a topographic map, the measurements are of elevation. Other isoline maps show such measurements as temperature, air pressure, soil type, or age of bedrock.

GRADIENT

You can use the information on a contour map to determine the average slope, or gradient, between two points on a landform. **Gradient** is defined as the change in elevation per unit distance. You can calculate the gradient between two points by using the following equation:

$$\text{Gradient} = \frac{\text{change in elevation}}{\text{distance}}$$

The change in elevation is determined by reading the elevation values on the contour map. The distance is determined by using the map scale. For example, suppose two locations are 5 kilometers apart. One is at an elevation of 2400 meters above sea level, and the other is at an elevation of 2600 meters above sea level. The gradient between the two locations is

$$\text{Gradient} = \frac{\text{change in elevation}}{\text{distance}}$$
$$= \frac{2600 \text{ m} - 2400 \text{ m}}{5 \text{ km}}$$
$$= \frac{200 \text{ m}}{5 \text{ km}} = 40 \text{ m/km}$$

In Chapter 1 you learned that one of the most important characteristics of science is that it can be presented in a manner that allows others to check your work. The calculation above shows each step so that it can be verified (or refuted) based on

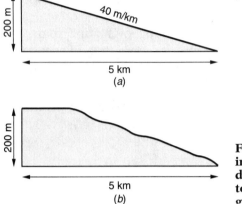

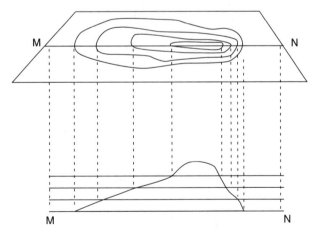

Figure 3-7. The slope of an incline is its gradient. The land does not have to be smooth to determine the average gradient.

the information in the calculation. Every time you use an algebraic equation to calculate an answer, you should present the solution in this form.

It is sometimes easier to study relief when you can see the landforms from the side rather than from above. A **profile** shows the relief across any part of a contour map. A profile can be made by drawing a line across a contour map. By plotting the

Figure 3-8. A profile can be made from a contour map.

points at which the line crosses the contour lines on the map, a vertical representation of the land is formed.

EARTH'S MAGNETIC FIELD

Maps are necessary for people who need to travel from one place to another. However, sailors navigated the seas long before accurate maps were created. How did they accomplish this? Early navigators relied on compasses. A **compass** is a device made up of a small bar magnet that is free to rotate. (You are familiar with magnets if you have one on your refrigerator or locker.) A bar magnet is a straight magnet that has two ends, called poles, where the magnetic effects are strongest. One end is called a north pole and the other a south pole. When two bar magnets are placed near each other, they interact. The north pole of one magnet will be attracted to the south pole of the other, and vice versa.

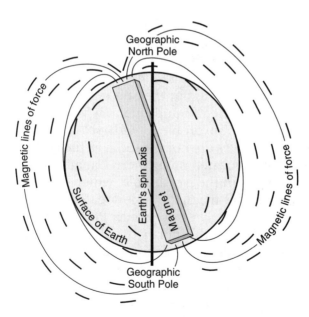

Figure 3-9. Earth's magnetic field resembles that of a huge bar magnet.

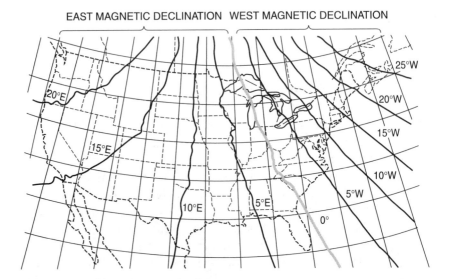

Figure 3-10. Magnetic declination for various locations in North America.

A compass works because Earth acts like a huge bar magnet. So the poles of the magnet in the compass are attracted to the magnetic poles of Earth. Scientists do not completely understand Earth's magnetism, but they do know that it is related to the movement of molten (liquid) metal within Earth's core.

There is only one problem with relying on a compass: Earth's magnetic poles are not the same as the geographic poles. The magnetic pole located in northern Canada is actually about 1500 kilometers from the North Pole, which is often called true north. Similarly, the magnetic pole located near the coast of Antarctica is not directly at the South Pole. The angle between geographic north and the north to which a compass needle points is known as **magnetic declination**. Magnetic declination varies with location. A navigator at sea or a hiker in the mountains must account for magnetic declination in order to use a compass. A person in central California, for example, would have to head about 15 to 20 degrees west of a compass reading to get to a place that is directly north on a map.

QUESTIONS

Multiple Choice

1. The imaginary line that circles Earth halfway between the poles is the *a.* equator   *b.* meridian   *c.* parallel *d.* prime meridian.

2. How many coordinates are needed to locate a position on Earth?   *a.* 1   *b.* 2   *c.* 3   *d.* 4

3. What is the latitude of the North Pole?   *a.* 0° North *b.* 90° North   *c.* 180° North   *d.* 360° North

4. For a person in the United States, latitude increases as you travel   *a.* north   *b.* south   *c.* east   *d.* west.

5. Every point on a contour line has the same   *a.* slope *b.* temperature   *c.* rock type   *d.* elevation.

6. Sailors who rely on a compass are using   *a.* Earth's magnetic field   *b.* Earth's shape   *c.* the sun   *d.* the moon.

Fill In

7. _Globe_ maps are the only type that represent Earth without distorting shapes or distances.

8. The _Slope_ on a topographic map represents the difference in elevation between two contour lines.

9. Elevation measurements are usually compared with the elevation at _distance_.

10. The coordinates for a point on Earth are its _latitude_ and longitude.

11. Longitude is the angular distance east or west of the _Prime meridian_.

12. Earth's magnetic field is thought to be a result of Earth's _molten liquid_.

Free Response

13. Why are different types of maps used?   *Cause one type of map is useful for one subject*

14. What are latitude and longitude?   *degrees*
    latitude and longitude are used for Plotting a location. latitude is formed around the equator at a 180° to make 360°

Gradient is one point to Another point
of travel

15. What is gradient and how is it calculated? *and direction*

16. Why does a compass needle point toward the north?
cause the magnet is the lava and liquid and metals are

17. What is magnetic declination? *1,500 kilometers west on*
Magnetic declination *North pole.*

Portfolio *15 to 20 degrees*
west of the north pole and East 15 to 20
or south.

Make an isoline map of the temperatures in your science classroom. Start by making a diagram of your classroom, showing all permanent pieces of furniture and fixtures along the floor. Mark rows of numbers along the floor up to about 20 or 30 locations to make a reasonably even coverage of the floor. (You can put them on the floor with masking tape.) Then use thermometers to record the temperature at each location.

Living Room

Temp: 60^F, 40^C

Flooring: Carpet Shag, Red Ruge.

Square Room: Left Right back Left Right.

2yft LR
4ft
24 x 24
Square
feet
480'

Chapter 4
Matter and Minerals

DIAMONDS AND EMERALDS are among the most beautiful gems on Earth, yet they can be grouped together along with rock salt and asbestos. What do these substances have in common? They are all minerals. They are also examples of matter. In this chapter, you will learn about the properties of matter and find out about the classification of minerals.

WHAT IS MATTER?

Earth and everything on it is made up of matter. **Matter** is anything that has mass and volume. Mass is the amount of material in an object. Volume is the amount of space an object takes up. You, your book, and your desk are all examples of matter.

Matter can be divided into two categories: mixtures and pure substances. A **mixture** contains two or more substances that are mixed together but not chemically combined. The substances in a mixture keep their individual properties. A mixture does not have definite proportions. In other words, different mixtures of the same substances can involve very different amounts of each. For example, you can form a mixture by adding sugar to iced tea. However, you might add a lot of sugar or just a little. Another mixture, ocean water, contains salt and water. However, different oceans, and even different locations within the same ocean, have different amounts of salt and water. Similarly, air is a mixture of several different substances, such as oxygen and nitrogen. The actual amounts of the various substances

can vary. Most mixtures can be separated into individual substances by physical means. For example, salt water can be separated into salt and water by allowing the water to evaporate.

Unlike a mixture, a **pure substance** is made of only one
kind of matter. Some pure substances are called elements. An
element is a substance that cannot be broken into simpler substances by ordinary chemical means. Hydrogen, oxygen, and helium are examples of common elements. There are about 100
different elements. If you look at the list of elements shown in
the Appendix, you will see that each element is represented by a
one- or two-letter symbol. Some symbols are taken from the first
letters of the element's name. Others are taken from their Latin
or Greek names.

Elements are composed of even smaller particles called
atoms. An **atom** is the smallest particle of an element that re-
tains the chemical identity of that element. Atoms are extremely
small. Even the largest atom is too small to be seen with an ordinary microscope. All of the atoms within any element are exactly alike. So, for example, gold contains only gold atoms, and
copper contains only copper atoms.

Other pure substances are compounds. A **compound** is a
substance that contains two or more elements that are chemically combined. A compound forms when the atoms within the
elements combine to form molecules. A **molecule** is the smallest part of a compound that still has all the properties of that
compound. The force that holds two atoms together is called a
chemical bond. Water, carbon dioxide, and table salt are examples of chemical compounds. Chemical compounds can be broken down into the elements from which they are formed. As in
the case of elements, symbols are used to represent compounds.
A combination of symbols used to show the elements in a compound is called a chemical formula. The formula for water, for
example, is H_2O. This tells us that a water molecule is composed
of two atoms of hydrogen and one atom of oxygen.

STATES OF MATTER

There are three states of matter: solid, liquid, and gas. A
solid has a definite volume and a definite shape. A marble is a

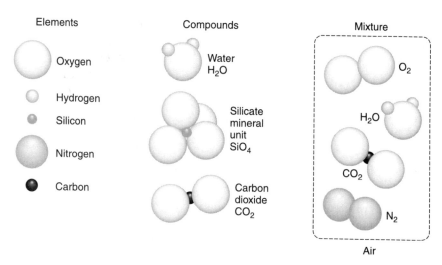

Figure 4-1. Elements, compounds, and mixtures are different types of matter.

solid object. It will keep its volume and shape no matter where you place it. The particles in a solid are packed closely together. Each particle is fixed in one position. The only motion of the particles is a slight vibration, or back-and-forth movement.

A **liquid** has a definite volume but not a definite shape. A liquid will take the shape of its container. You can pour 100 milliliters of water into a round vase or into a beaker. The volume of water, 100 milliliters, will not change. The reason is that although the particles in a liquid are packed almost as closely as in a solid, they are free to move around one another.

Unlike a solid or liquid, a **gas** does not have a definite volume or a definite shape. The particles in a gas move around at high speeds in all directions. The particles will spread out to fill a large container. They can also be pushed together in a small container. The air you breathe, the helium in a balloon, and the neon in a light are all substances in the gaseous state.

The same substance can exist in more than one state of matter. You are already familiar with this fact if you use ice cubes to cool a glass of water. Water can exist as solid ice, liquid water, or gaseous water vapor. What determines the state of matter? The answer is energy. The more energy a particle of matter has, the

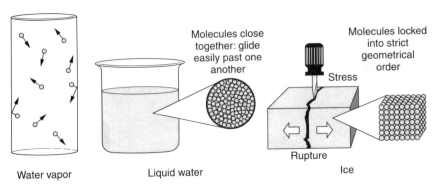

Figure 4-2. Matter, such as water, can exist in three states.

faster and farther it moves. The particles in a solid tend to have less energy than those in a liquid. The particles in a liquid have less energy than those in a gas. When energy is added or removed, a substance can change from one state of matter to another. You can add or remove energy by heating or cooling a substance. As you heat solid ice, it turns to water and eventually to steam. The substance, water, does not change—only its state of matter changes.

WHAT IS A MINERAL?

As you study earth science, you will encounter many different types of matter. In this chapter, and the following two chapters, you will learn about matter in the form of rocks and minerals. A **mineral** is a substance that is solid, has a definite chemical composition, occurs naturally, is inorganic, and has a crystal structure. To help you identify minerals, let's look at the properties of minerals more closely.

- All minerals are solids; they have a definite shape and volume. Gases, such as hydrogen, cannot be minerals.
- Minerals have a definite chemical composition. This means that the elements in most minerals are found in proportions that do not change. Some minerals are mixtures. Olivine is a mixture of iron silicate and magnesium silicate.

- A mineral must occur naturally on Earth. Silver and gold, which are minerals, occur naturally on Earth. Plastics, which are manufactured, do not and are therefore not minerals.
- A mineral must be inorganic. An object is inorganic if it was not formed by living things. Coal and oil are not minerals because they were formed from the remains of living things.
- The atoms in a mineral are arranged in a definite pattern that is repeated. This repeating pattern forms a solid **crystal**, which is a geometric shape with flat sides.

Some minerals are very common, whereas others are rare and valuable. In either case, all minerals share the five characteristics just described.

There are more than 2000 different kinds of minerals. Familiar minerals include feldspar, quartz, mica, gold, and diamond. Most of these minerals are compounds. Quartz, for example, is a mineral made up of silicon and oxygen atoms (SiO_2). Quartz is an extremely common mineral that is abundant in sand and a wide variety of rocks. Calcite, the primary mineral in limestone, is a mineral made up of calcium, carbon, and oxygen ($CaCO_3$). A few minerals are made up of single elements. A mineral composed of only one element is called a **native element**. Examples of native elements are gold (Au), copper (Cu), silver (Ag), and sulfur (S). Graphite and diamond, two forms of carbon (C), are also native elements.

Most rocks in Earth's crust are made of minerals. Some rocks, such as limestone and shale, are composed primarily of a single mineral. (In these two cases, calcite and clay respectively.) But most rocks contain a variety of minerals. For example, granite, a relatively common type of rock with visible mineral crystals, usually contains at least four different minerals. Other minerals may or may not be found in granite, but they are generally present in smaller portions. Common minerals that make up most of the rocks in Earth's crust are called rock-forming minerals. Quartz, calcite, micas, hematite, augite, and feldspar are examples of rock-forming minerals.

Of the more than 100 elements, eight make up over 98 percent of the mass of Earth's crust. More than 90 percent of the

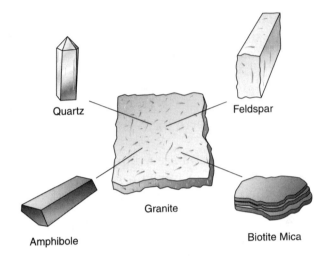

Figure 4-3. Most rocks, such as this granite, are composed of a number of minerals.

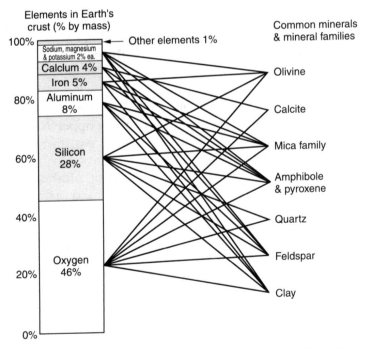

Figure 4-4. These eight elements and eight minerals make up the majority of rocks around us. Follow the lines to find out which elements are in each type of mineral.

minerals in Earth's crust are silicates. **Silicates** are minerals that contain silicon and oxygen, along with one or more metals such as aluminum or iron.

Where do minerals come from? Many minerals form from molten rock, or magma, which is rock that is so hot it is in the liquid state. When magma cools and hardens, mineral crystals may form. Some other minerals form from compounds dissolved in liquids such as water. When the liquid evaporates, or changes from a liquid to a gas, it leaves behind mineral crystals. The mineral halite, for example, forms when salt water evaporates.

QUESTIONS

Multiple Choice

1. A substance that contains two or more elements that are chemically combined is called   *a.* an atom   *b.* a mixture   *c.* an element   *d.* a compound.
2. The state in which matter has a definite shape and volume is   *a.* solid   *b.* liquid   *c.* gas   *d.* magma.
3. Which is not a property of minerals?   *a.* solid   *b.* organic   *c.* naturally occurring   *d.* crystal structure
4. Gold and silver are examples of   *a.* silicates   *b.* native minerals   *c.* compounds   *d.* organic substances.
5. The most common element by mass found in rocks near Earth's surface is   *a.* silicon   *b.* iron   *c.* oxygen   *d.* aluminum.

Fill In

6. _____ is anything that has mass and volume.
7. Matter can be divided into _____ and pure substances.
8. The three states of matter on Earth are solid, liquid, and _____.
9. Inorganic solids called _____ make up nearly all rocks on Earth's surface.
10. Most of the mass of Earth's crust is made up of minerals called _____.

11. Many minerals are formed from _____, which is rock in the liquid state.

Free Response

12. What is matter?
13. Why is air classified as a mixture?
14. Describe the three states of water in terms of motion of water molecules.
15. What is a mineral?
16. Explain why ice can be classified as a mineral, but water from the faucet cannot.
17. Coal is a rock that is not composed of minerals. Why?

Portfolio

Find out about common minerals. Make a list of minerals that are useful to people and explain how they are used. If possible, include photographs from magazines or the Internet to show examples of minerals and their uses.

Chapter 5

Mineral Identification

YOU SEE A GLITTERING YELLOW SUBSTANCE in the ground. Is it gold? Not necessarily. It might be a mineral called pyrite, which is also known as "fool's gold." There are so many different minerals, it can be difficult to tell them apart unless you know what to look for. Minerals are identified by their physical properties, which you can observe by looking at the mineral or by conducting simple physical tests. The study of minerals and their properties is known as mineralogy.

PROPERTIES OF MINERALS

The properties that geologists find most useful for identifying minerals include color, luster, crystal form, hardness, streak, cleavage, and specific gravity.

Color

Color is the property of minerals that is easiest to observe. This property is useful in identifying minerals that have their own characteristic colors. For example, malachite is green, azurite is blue, and garnet is deep red. However, color may be a misleading property for identifying some minerals because different minerals can have similar colors. Calcite, feldspar, and several other minerals are all milky-white in color. In addition, small amounts of impurities can change the color of a mineral. Pure quartz, for example, is white or colorless. Depending on its impurities, however, quartz can also be pink, green, gray, or nearly any other color. Dark-colored minerals such as the amphiboles are less likely to show variations in color from sample to sample, because impurities are unlikly to affect their color.

Luster

The way that a mineral reflects light from its surface is called **luster.** Luster can be metallic or nonmetallic. A mineral with a metallic luster reflects light much like a highly polished metal does. Such minerals include silver, copper, gold, galena, and pyrite. Minerals that do not shine like metals have a nonmetallic luster. A variety of different names is used to describe minerals with nonmetallic lusters, such as glassy, brilliant, pearly, waxy, dull, and greasy. Do not confuse luster with shininess. Glass, for example, can shine in light, but it has a nonmetallic luster because light penetrates its surface.

Crystal Form

As you learned in Chapter 4, one property of minerals is their crystal form. Crystals are sometimes called the "flowers" of the mineral world because they often have regular geometric shapes and beautiful colors. The shape of a mineral crystal is a result of the internal arrangement of its atoms. Crystals have flat faces, sharp edges, and geometric symmetry. Quartz forms six-sided (hexagonal) crystals. Calcite crystals look like rectangular solids that have been pushed over toward one corner.

In nature, perfect crystals are hard to find. Crystal faces can form only when minerals have enough room and time to grow large. Yet the mineral grains in most rocks do not have enough room to form perfect crystal faces.

Hardness

The **hardness** of a mineral is its resistance to being scratched. Any mineral can scratch another mineral that is softer than itself, but it cannot scratch a mineral that is harder than it is. To give a specific measure to hardness, the mineralogist

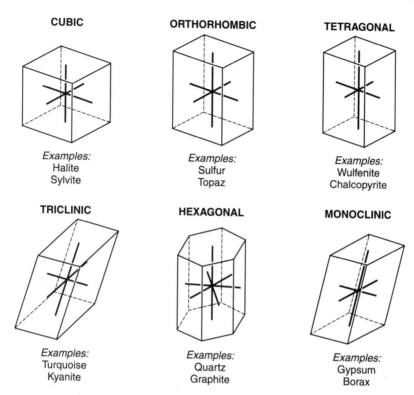

CUBIC

Examples:
Halite
Sylvite

ORTHORHOMBIC

Examples:
Sulfur
Topaz

TETRAGONAL

Examples:
Wulfenite
Chalcopyrite

TRICLINIC

Examples:
Turquoise
Kyanite

HEXAGONAL

Examples:
Quartz
Graphite

MONOCLINIC

Examples:
Gypsum
Borax

Figure 5-1. Six common crystal types.

Friederich Mohs devised a hardness scale. The scale, known as Mohs' scale, arranges ten familiar minerals in order of hardness from 1 to 10. The softest mineral on Mohs' scale is talc, which has a hardness of 1. All other minerals on this scale can scratch talc. The hardest natural substance is diamond, which has a hardness of 10. No other mineral can scratch a diamond, but a diamond can scratch any other natural substance. Feldspar and quartz, very common minerals, have hardness values of 6 and 7 respectively on Mohs' scale. This means that quartz can scratch feldspar, but feldspar cannot scratch quartz.

You can use Mohs' scale to find the approximate hardness of any common mineral by rubbing it against minerals from the scale. For example, suppose an unknown mineral scratches talc, gypsum, and calcite but not fluorite. If fluorite scratches the unknown mineral, you know that the unknown mineral has a hard-

Table 5-1. Mohs' Scale of Hardness

Hardness	Mineral	Common Objects for Comparison
1	Talc	Fingernail (2.5) scratches it easily.
2	Gypsum	Fingernail (2.5) scratches it.
3	Calcite	Copper penny (3.5) just scratches it.
4	Fluorite	Iron nail (4.5) scratches it.
5	Apatite	Glass (5.5) scratches it.
6	Feldspar	It scratches glass.
7	Quartz	It scratches steel and glass easily.
8	Topaz	
9	Corundum	
10	Diamond	

ness between 3 and 4. If fluorite does not scratch the unknown mineral, it must have a hardness of 4 because two minerals that do not scratch each other have the same hardness.

Streak

If you rub a mineral against a hard, rough surface, some powder is scraped off the mineral. The color of the powder of a mineral is called **streak.** Streak is tested with a special white porcelain tile called a streak plate. A streak plate is much like the bottom of a bathroom or floor tile. When a mineral sample is rubbed on a streak plate, the mark left behind shows the color of the mineral powder. Although the color of a mineral may vary, its streak is always the same color. For example, hematite may be brown, red, or silver, but its streak is always reddish-brown. This property is useful in identifying minerals with a metallic luster. The streak of a metallic mineral is at least as dark as the mineral itself. The streak test is not as useful in identifying non-metallic minerals because most nonmetallic minerals leave a white or colorless streak.

Cleavage

The term **cleavage** refers to the tendency of a mineral to split along smooth, flat surfaces. Cleavage surfaces can be recognized by the way they reflect light, much like a polished, flat

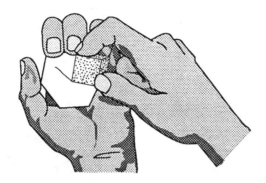

Figure 5-2. The streak test is performed by rubbing a clean edge of the mineral across a white streak plate.

surface. Some minerals, such as the mica family, split (or cleave) into thin, flat sheets. Mica minerals are said to show perfect cleavage in one direction. The mineral feldspar commonly cleaves in two directions that are at nearly right angles (90°). Feldspar is said to have two good cleavages. Halite (rock salt) splits into cubes and rectangular solids with three perpendicular cleavage planes. Halite is said to have three 90° cleavages.

Although cleavage is a useful property for mineral identification, not all minerals have cleavage. Other minerals can break, but they do so along irregular surfaces. Quartz, for example, breaks along curved or irregular surfaces. Minerals that break along irregular surfaces are said to have **fracture**.

Specific Gravity

All matter can be described in terms of density. The density of a substance is its concentration of matter. More precisely, **density** is mass per unit volume, or grams per cubic centimeter. Cork is less dense than steel, and oil is less dense than water. If two objects are the same size, the denser object will be heavier. And an object will float in a liquid if the object is less dense than the liquid. The mineral with the greatest density is gold. Its density is 19.3 g/cm^3.

Mineralogists can draw conclusions about an unknown mineral by comparing the density of a sample with the density of a

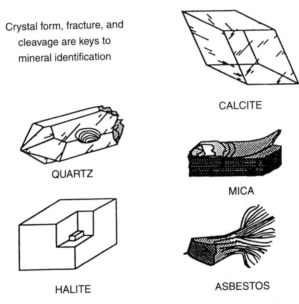

Crystal form, fracture, and
cleavage are keys to
mineral identification

CALCITE

QUARTZ

MICA

HALITE

ASBESTOS

Figure 5-3. Some minerals split along flat surfaces
(cleave) when they are broken, others fracture.

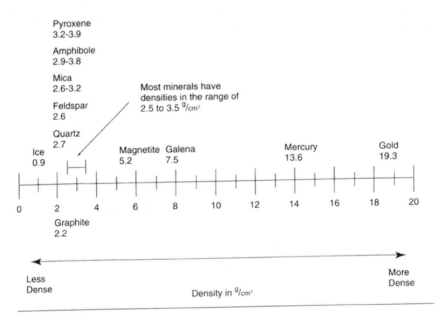

Pyroxene
3.2-3.9

Amphibole
2.9-3.8

Mica
2.6-3.2

Feldspar
2.6

Quartz
2.7

Most minerals have
densities in the range of
2.5 to 3.5 $^g/cm^3$

Ice
0.9

Magnetite   Galena
5.2   7.5

Mercury
13.6

Gold
19.3

0   2   4   6   8   10   12   14   16   18   20

Graphite
2.2

Less
Dense

More
Dense

Density in $^g/cm^3$

Figure 5-4. Many common minerals have densities between 2.5 and
3.5 g/cm^3.

known substance, such as water. A comparison of the density of a substance with the density of water is known as **specific gravity**.

$$\text{Specific gravity} = \frac{\text{Density of mineral (g/cm}^3)}{\text{Density of water (g/cm}^3)}$$

The density of water is 1.0 g/cm^3. Notice that when the calculation is completed, the units of density cancel out. Therefore, a specific gravity measurement has no units. So the density of quartz is 2.7 g/cm^3, but the specific gravity of quartz is simply 2.7.

Special Properties

Some minerals have properties that set them apart from nearly all others. Although testing for these specific properties may be useful for only a single mineral, these properties can provide definite identification.

The mineral calcite has two special properties. One property is known as **double refraction**, which means that it splits light rays into two parts. The result is that a thin line viewed through a transparent calcite crystal appears as two parallel lines. The other property can be identified through a chemical test. When a strong acid is placed on a sample of calcite, the mineral fizzes as carbon dioxide is released.

The mineral halite (rock salt) is special because of its salty **taste**. (CAUTION: In general, you should not taste substances in a science laboratory. However, your teacher may allow you to use the taste test in some rock and mineral activities. Be sure to get permission first.)

Some minerals, such as magnetite, are attracted to a magnet. Some other minerals, such as the uranium minerals carnotite and uraninite, are radioactive. This means that they release subatomic particles. This property can be tested with a Geiger counter. Other minerals show fluorescence. That is, they glow when they are placed in ultraviolet light.

The best way to become skilled at mineral identification is to observe and identify minerals using a wide variety of relatively pure mineral samples. Until you become proficient, avoid weathered rocks or rocks in which the individual minerals are

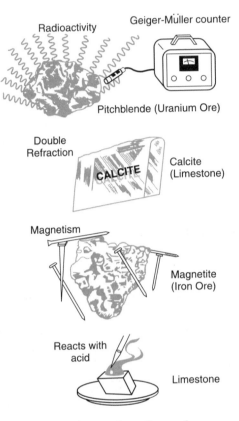

**Figure 5-5. Some minerals can be
identified by their unusual properties.**

difficult to distinguish. The flowchart in Figure 5-6 can help you
identify the most common minerals through their properties.
Less common minerals may require more specialized information and tests.

THE MOST COMMON MINERALS

So far in your study of minerals, you have come across the
names of several common minerals, such as quartz, feldspar,
and mica. Before going any farther, let's look at these common
minerals in a little more detail.

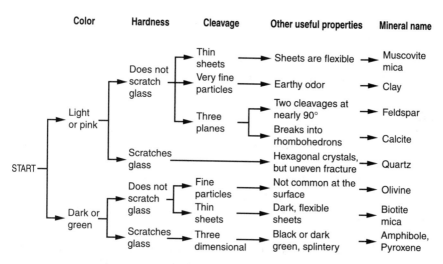

Figure 5-6. This flowchart shows the steps to identifying the most common minerals. Start at the left side of the chart and progress through the properties of the sample.

Feldspar

In Chapter 4 you learned that the most common minerals are silicates, which means that they contain silicon and oxygen in addition to other elements. One family of silicates, the feldspars, accounts for more than 60 percent of Earth's crust. Different types of feldspar include those that contain potassium, such as orthoclase, and those that contain sodium-calcium, such as albite and oligoclase. All feldspar minerals break into pieces with two smooth cleavage surfaces at nearly right angles, have a pearly luster, and have a hardness of 6 on Mohs' scale.

Quartz

Quartz is the second most abundant mineral in Earth's crust. Unlike the feldspars, quartz is made entirely of silicon and oxygen. Quartz is a relatively hard mineral with a hardness of 7 on Mohs' scale. It is usually light in color and has a glassy luster. Pure quartz can be as transparent as glass, but it is more often translucent. (If an object is translucent, some light passes through it but not enough for you to see objects clearly through it.) Perfectly formed quartz crystals are hexagonal (six-sided) and usually come to a point at one end. Quartz breaks along uneven surfaces. Quartz exists in many forms, including the semi-precious amethyst, agate, and onyx.

Mica

The micas make up a group of soft silicates. Mica is easy to identify because it is the only common mineral that breaks into thin, flexible sheets. Mica has one perfect cleavage. One type of mica, biotite, is very dark in color due to the presence of iron and magnesium. Another type of mica, muscovite, is silvery white. Muscovite, which consists primarily of silicate of potassium and aluminum, sometimes lends a silvery sheen to rocks. These minerals have a hardness of about 2.5 on Mohs' scale.

Amphibole and Pyroxene

Both amphibole and pyroxene are common families of dark-colored minerals that can break into stubby splinters. These minerals have a hardness between 5 and 6. The best way to tell them apart is the angle between cleavage faces. Amphiboles, such as hornblende, form cleavage surfaces that meet at nonperpendicular angles such as 60° or 120°. Members of the pyroxene family of minerals, such as augite, cleave at right angles (90°).

Calcite

Calcite is a common mineral that is not a silicate. Instead, calcite is known as a carbonate because it contains a carbonate group, which is a carbon atom bonded to three oxygen atoms. Calcite, which consists of a calcium atom bonded to a carbonate group, is the most common carbonate mineral. It may look clear and glassy, but like quartz it can be found in almost any color. Calcite can be scratched with a knife blade or a steel nail because it has a hardness of 3 on Mohs' scale. If broken, calcite has three perfect cleavages. It forms crystals that look like a cube pushed over toward one corner. Calcite is the major mineral in limestone. The stalagmites and stalactites in caves are usually formed from calcite.

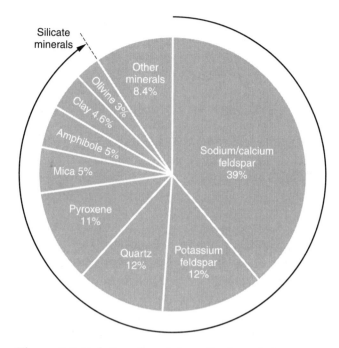

Figure 5-7. Relative abundance of minerals in Earth's crust.

QUESTIONS

Multiple Choice

1. The scratch test is used to determine *a.* color *b.* hardness *c.* density *d.* luster.

2. Nonmetallic is a classification of *a.* color *b.* hardness *c.* density *d.* luster.

3. Which test is used to find the color of the powder of a mineral? *a.* hardness *b.* specific gravity *c.* luster *d.* streak

4. Which are the metric units of measure for density? *a.* grams *b.* cubic centimeters *c.* grams per cubic centimeter *d.* cubic centimeters per gram

5. A single line will look like two lines when viewed through a clear crystal of *a.* quartz *b.* calcite *c.* feldspar *d.* pyrite.

6. The mineral with the highest density is   *a.* quartz   *b.* feldspar   *c.* calcite   *d.* gold.

Fill In

7. The _____ of a mineral is the way it reflects light.

8. The softest mineral on Mohs' scale is _____.

9. Although the observable color of a mineral may vary, the color of its _____ is always the same.

10. The property known as _____ is the tendency of a mineral to split easily.

11. The _____ of water is 1 g/cm^3.

Free Response

12. List the six basic properties that can be used to identify minerals. Explain how you might divide the minerals into two groups based on each property.

13. Why can't you choose one of the properties of minerals and use it to identify a sample of unknown minerals?

14. Describe the two types of luster and give examples of each.

15. What is the purpose of Mohs' scale?

16. What is the difference between cleavage and fracture?

17. How is specific gravity related to density?

18. How are quartz and feldspar alike and how are they different?

Portfolio

I. Make a collection of local minerals. Try to find clean, fresh samples that show the mineral's characteristic properties. Prepare a poster that shows how you classified each mineral.

II. Mineral samples are largely dependent on geographical location. Trade mineral samples with students in another school that is in different kind of geological setting.

Chapter 6
The Formation of Rocks

You can probably spot several rocks just by looking out your window. Rocks come in various sizes, shapes, and colors. They form all of Earth's crust from the tops of the highest mountains to the bottom of the ocean floor. A **rock** can generally be defined as a solid substance composed of one or more minerals. You can think of minerals as the "building blocks" of rocks. Rocks are classified into three groups according to how they were formed: igneous, sedimentary, and metamorphic. In this chapter, you will learn about the different types of rocks and how they change from one type into another.

IGNEOUS ROCKS

You are familiar with water in both its solid state and its liquid state. Do you think rock has a liquid state as well? You may be surprised that the answer is yes. At high temperatures and pressures, rock can exist in a liquid, or molten, state. You rarely see rock in this form because the conditions required to melt rock are usually limited to locations deep within Earth. Rocks that are formed when hot molten rock cools and hardens are known as **igneous** rocks.

Molten rock can harden in two ways. One way is through a slow process deep within Earth. Molten rock underground is known as **magma**. Rocks formed deep undergound are called the **plutonic**, or **intrusive**, igneous rocks. Plutonic rocks are found on Earth's surface only after the rock covering them has

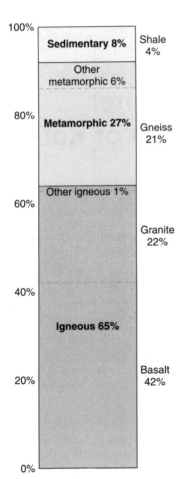

Figure 6-1. Abundance of igneous, metamorphic, and sedimentary rocks in Earth's crust.

worn away over time. Granite is a common intrusive igneous rock that is sometimes used as a building stone.

Igneous rock also forms when magma is pushed up and onto Earth's surface during a volcanic eruption. Magma that pours onto Earth's surface is known as **lava**. The igneous rocks that form from lava cool quickly. Igneous rocks formed in this way are known as **volcanic**, or **extrusive**, igneous rocks.

Mineral crystals grow over time. The longer they have to develop, the larger they will be. The size, shape, and arrangement of the mineral crystals within a rock determine the **texture** of the rock. There are four common igneous textures. In order of in-

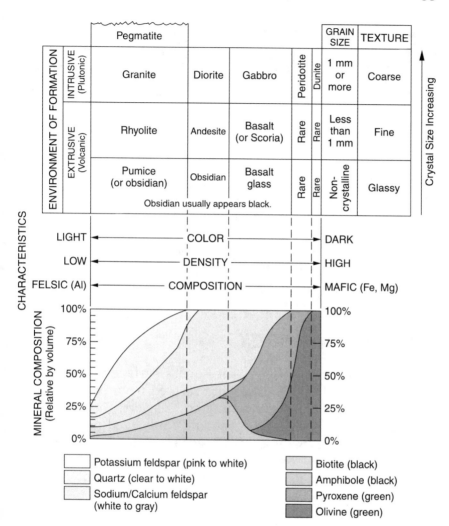

Figure 6-2. Scheme for igneous rock identification.

creasing crystal size they are glassy-smooth, frothy, fine-grained, and coarse-grained. The crystals in intrusive igneous rocks are large because the magma takes a long time to harden. In fact, the mineral crystals in intrusive rocks are large enough to be clearly visible without magnification. Thus intrusive igneous rocks, such as granite, tend to have more coarse-grained textures because the crystals are large. Extrusive igneous rocks form quickly. Since

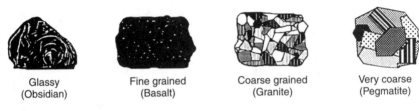

| Glassy (Obsidian) | Fine grained (Basalt) | Coarse grained (Granite) | Very coarse (Pegmatite) |

Figure 6-3. Four representative igneous rocks.

the mineral crystals within the rocks do not have much time to grow, they remain small and are not readily visible. Thus extrusive igneous rocks, such as basalt, tend to have fine-grained textures because the crystals are small. In some cases, the lava hardens so quickly that the crystals barely have time to develop at all. The rocks that form under these conditions are a natural glass called obsidian and a frothy form called pumice.

SEDIMENTARY ROCKS

The most common rocks on Earth's surface are sedimentary rocks. **Sedimentary** rocks are formed from particles that have been deposited by water, wind, glaciers, or landslides. These particles, or sediments, become compacted or cemented together over time. Sedimentary rocks form at or near the surface of Earth. They usually are easy to recognize because the sediments are deposited in layers. The arrangement of layers is known as **stratification**.

Any rock that contains fossils is almost certainly a sedimentary rock. **Fossils** are the remains or any other evidence of living organisms, such as plants and animals, preserved in rock. Fossils are found in sedimentary rocks because when an organism dies, it can become buried in loose sediments. Although the soft parts of the organisms decay, the hard parts may be preserved as fossils. For example, the shells of clams and mussels are often found in sandstone, limestone, and shale. Even if the entire organism decays, an impression might be left in the rock. Plant and animal impressions can be seen when the rock is split apart. Sedimentary rocks can be classified as fragmental, organic, or chemical.

INORGANIC LAND-DERIVED SEDIMENTARY ROCKS					
TEXTURE	GRAIN SIZE	COMPOSITION	COMMENTS	ROCK NAME	MAP SYMBOL
Clastic (fragmental)	Sand, pebbles, cobbles, boulders	Mostly quartz, feldspar, clay minerals	Particles rounded and cemented by fine particles	Conglomerate	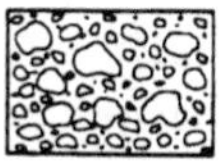
	Sand		Can be fine to coarse	Sandstone	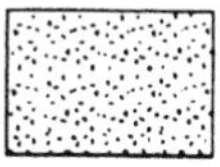
	Silt		Can be compact or easily split	Siltstone	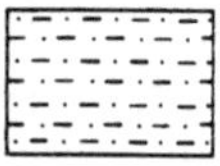
	Clay			Shale	

CHEMICALLY AND ORGANICALLY FORMED SEDIMENTARY ROCKS					
TEXTURE	GRAIN SIZE	COMPOSITION	COMMENTS	ROCK NAME	MAP SYMBOL
Nonclastic	Coarse to fine	Calcite	Crystals from chemical precipitates (Incudes the evaporites)	Chemical Limestone	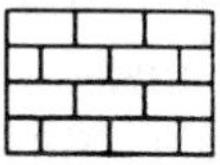
	All sizes	Mostly halite		Rock Salt	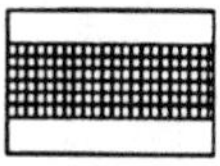
	All sizes	Gypsum		Rock Gypsum	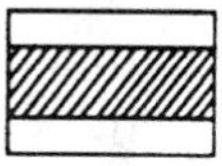
	All sizes	Dolomite	Changed from limestone by replacement	Dolostone	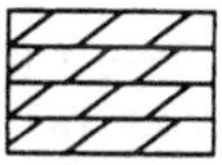
Organic	Microscopic to coarse (larger than 0.2 cm)	Calcite	Cemented shells, shell fragments, and skeletal remains	Fossil Limestone	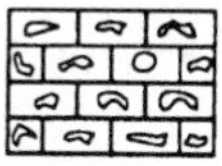
	All sizes	Carbon from plant remains	Black and nonporous	Coal	

Figure 6-4. Scheme for sedimentary rock identification.

Fragmental

The most common sedimentary rocks are **fragmental**, or **clastic**, rocks, which are formed from the weathered fragments of other rocks. The fragments are then carried by natural processes, such as running water. What makes the particles stick together? Water in oceans, lakes, and the ground contains substances that act as cement to hold the sediment together. These natural cements, including silica, calcite, and iron oxide, bind the fragments together into rock. The color of the rock is related to the cement. Silica or calcite causes the rocks to be gray or white. Iron oxide creates red or brown rocks.

The properties of fragmental rocks depend on the size of the sediment from which they are formed. Sediment ranges in size from large pebbles to microscopic flakes. Clay, for example, is very fine sediment that is abundant in the most common type of sedimentary rock, shale. A fresh sample of shale feels smooth and soft. Sandstone is composed of larger grains of sediment, which make it feel rough and gritty. Conglomerate contains the largest particles, including pebbles or even larger particles. This makes conglomerate the coarsest of the fragmental rocks.

Organic

Organic sedimentary rocks are formed from plant or animal remains. Coal and limestone are common rocks that come from organic sediments. Coal, for instance, is formed when the remains of plants are changed under pressure over millions of years. Limestone often forms from the shells of animals such as clams, mussels, oysters, and corals.

Chemical

A mixture in which one substance is dissolved in another is known as a solution. Many natural supplies of water, such as oceans, lakes, and underground waters, are solutions of water that contains dissolved minerals. Chemical sediments are formed when the minerals come out of solution. One way this happens is for the water to evaporate, leaving behind the solid minerals. Rock salt is a mineral left behind when seawater evaporates.

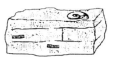

| Fine-grained and layered (shale) | Layers of gritty particles (sandstone) | Cemented pebbles (conglomerate) | Thick layers with fossils (limestone) |

Figure 6-5. Four representative sedimentary rocks.

METAMORPHIC ROCKS

Metamorphic rocks result from dramatic changes in existing rocks caused by heat and/or pressure. Metamorphic rock is the only kind that always forms directly from another rock. Marble, slate, gneiss, and quartzite are metamorphic rocks.

The process of metamorphism causes a variety of changes. The rock is compressed. In other words, the rock becomes more dense. In addition, heat and chemicals can rearrange the particles within the rock, causing new minerals to be formed. Original structures, such as sedimentary layers or fossils, may become distorted or even destroyed. Metamorphism may result in features known as foliation or banding. **Foliation** is a fine layering caused when pressure squeezes mineral crystals into layers. Metamorphic rocks split easily along these layers. Foliation is different from stratification, which is layering caused as sediment is deposited in sedimentary rocks. **Banding** is the separation of minerals into light and dark layers. Banding usually occurs in metamorphic rocks such as gneiss that have experienced high heat and pressure deep within Earth.

There are two common types of metamorphism: contact metamorphism and regional metamorphism. **Contact metamorphism** occurs when molten magma moves up through the rock that lies over it. The introduction of magma into the rock is known as an intrusion. The rock closest to the intrusion is heated by the magma. In addition, substances from the magma enter the rock and react with its minerals. The changes in the rock decrease as distance from the magma increases. In other words, contact metamorphism is usually localized at or near the boundaries of an intrusion. An example of a rock formed by contact metamorphism is hornfels. Most metamorphic rock is formed by regional metamorphism. **Regional metamorphism** occurs when large

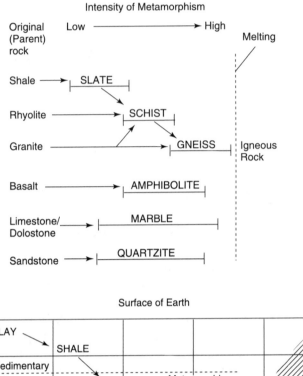

Intensity of Metamorphism

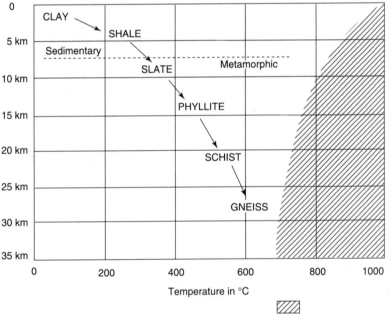

Conditions necessary to make various kinds of
metamorphic rocks, or to make magma, which
will cool to become igneous rock.

At these temperatures
rocks melt to become
igneous rocks.

**Figure 6-6. This chart shows the most common varieties of metamorphic
rocks and the original rock before metamorphism. The graph indicates
the conditions required to change rocks from one form to another.**

TEXTURE		GRAIN SIZE	COMPOSITION	TYPE OF METAMORPHISM	COMMENTS	ROCK NAME	MAP SYMBOL
FOLIATED	Slaty	Fine	CHLORITE / MICA / QUARTZ / FELDSPAR / AMPHIBOLE / GARNET / PYROXENE	Regional (Heat and pressure increase with depth, folding, and faulting)	Low-grade metamorphism of shale	Slate	
	Schistose	Medium to coarse			Medium-grade metamorphism; mica crystals visible from metamorphism of feldspars and clay minerals	Schist	
	Gneissic	Coarse			High-grade metamorphism; mica has changed to feldspar	Gneiss	
NONFOLIATED		Fine	Carbonaceous		Metamorphism of plant remains and bituminous coal	Anthracite coal	
		Coarse	Depends on conglomerate composition	Thermal (including contact) or regional	Pebbles may be distorted or stretched; often breaks through pebbles	Meta-conglomerate	
		Fine to coarse	Quartz		Metamorphism of sandstone	Quartzite	
		Fine to coarse	Calcite, dolomite		Metamorphism of limestone or dolostone	Marble	
		Fine	Quartz, plagioclase	Contact	Metamorphism of various rocks by contact with magma or lava	Hornfels	

Figure 6-7. Scheme for metamorphic rock identification.

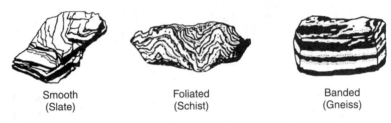

Smooth   Foliated   Banded
(Slate)   (Schist)   (Gneiss)

Figure 6-8. Three representative metamorphic rocks.

areas of rock are under intense heat and pressure that cause the rock to change form.

Given the parent rocks and the metamorphic conditions, the resulting rocks are predictable. For example, limestone changes into marble, and sandstone changes into quartzite. Shale can progress through a series of changes from slate to schist and finally to gneiss.

THE ROCK CYCLE

Planet Earth is basically a closed system. That means that Earth receives very little matter from outer space (mostly in the form of meteorites and dust) and loses very little matter into space (mostly the lighter atmospheric gases such as hydrogen). As a result, new rocks are invariably created from the remains of other rocks. The solid materials of Earth circulate through various forms shown in a dynamic model known as the **rock cycle**. It is a cycle because the same kinds of changes can occur over and over again through geologic time.

The rock cycle may be a little confusing to follow at first because of the various arrows. The arrows show possible changes that may occur. For example, find igneous rock in Figure 6-9. If you follow the arrows, you see that weathering and erosion break igneous rocks into sediments over time. These sediments are then formed into sedimentary rocks. If these sedimentary rocks are then exposed to conditions of high pressure and temperature, they will be changed into metamorphic rocks.

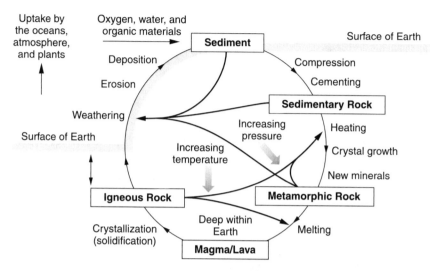

Figure 6-9. The rock cycle is a model that shows how rock materials change over time.

There are a variety of paths and shortcuts through the rock cycle because there is no single process progression for a given rock. Rather than being weathered to form sediments, for example, igneous rocks might be forced back into Earth's crust where they will melt back into magma. Or igneous rocks could be changed directly into metamorphic rock. Follow the arrows along the outside and inside of the circle to identify the many processes that can occur. Igneous and metamorphic rocks generally form near volcanoes or deep within Earth. These rocks are often exposed when uplift and erosion remove the soil and sedimentary rocks that cover them.

CHANGE AND CYCLES

Change, such as in the rock cycle, is the natural state of our planet. In fact, it is difficult to think of something that does not change. Some changes, such as the flash of lightning, are so fast they seem instantaneous. Other changes, such as the erosion of a mountain, occur so slowly they are difficult to observe in a lifetime. Change that always proceeds from start to finish is known as noncyclic change. The rusting of iron is an example of a

noncyclic change because the iron rusts until it is gone. A change that repeats itself is a **cyclic change**. The progression of the seasons of the year is an example of a cyclic change because the change repeats year after year. A cyclic change can be described by a model, such as the rock cycle. Another cycle in nature is the water cycle. In the water cycle, water falls to Earth as precipitation (rain, snow, and hail). Once on Earth, water forms lakes, rivers, and streams or sinks into the ground. Either way, the water eventually evaporates to form clouds. Then the cycle begins again. Can you think of other cycles? As you study later chapters, you will be reminded that this planet is changing constantly.

QUESTIONS

Multiple Choice

1. A rock that forms when magma cools and hardens is
   a. igneous   *b.* sedimentary   *c.* metamorphic   *d.* lava.

2. An example of an intrusive igneous rock is   *a.* sandstone
   b. limestone   *c.* granite   *d.* marble.

3. Fossils are usually found in   *a.* igneous rocks   *b.* sedimentary rocks   *c.* metamorphic rocks   *d.* lava rocks.

4. The evaporation of seawater may result in   *a.* igneous rocks   *b.* sedimentary rocks   *c.* metamorphic rocks
   d. organic remains.

5. The type of rock most common at Earth's surface is
   a. igneous   *b.* sedimentary   *c.* metamorphic   *d.* magma.

Fill In

6. A(an) _____ is a group of minerals bound together.

7. Magma is called _____ when it rises above the ground.

8. _____ rocks are formed when molten rock cools and hardens.

9. _____ rocks can be identified by stratification.

10. Marble is an example of a(n) _____ rock.

Free Response

11. List and describe the characteristics of the three different types of rocks.
12. How are intrusive rocks similar to extrusive rocks? How are they different?
13. Describe the changes that occur in the rock cycle.
14. How is a cyclic change different from a noncyclic change?

Portfolio

I. Prepare a collection of local rocks. Try to find samples that are fresh and unweathered. Classify them as igneous, sedimentary, or metamorphic. Give reasons for your classification.

II. Compose a written or video report on the composition, origin, and uses of a particular kind of rock, such as shale or granite.

III. Prepare an enlarged display of the rock cycle. Use photographs or samples of rocks for each of the rock types named on the cycle.

Chapter 7
Earth's Natural Resources

No MATTER HOW FAR society advances, everything people need to survive must come from planet Earth. You may find this hard to believe because you can buy your food, clothing, and other necessities in a store. However, the ultimate source of all these items is Earth. For example, Earth supplies the soil in which your vegetables are grown, the copper used to wire your school, and the aluminum to make your soft-drink cans. With the exception of a few materials, people and other animals are totally dependent on Earth's natural resources. A **natural resource** is a material supplied to living things by the environment. Earth's natural resources can be divided into two groups: nonrenewable and renewable. A **nonrenewable resource** exists in a fixed amount and may eventually run out. A **renewable resource** can be replaced in nature at a rate that is close to the rate at which it is used. In this chapter, you will learn about the different types of resources you use every day and what you can do to protect their supplies.

NONRENEWABLE RESOURCES

The minerals you have been reading about in the last several chapters are nonrenewable resources. Recall that minerals are found in rocks. Rocks that contain useful minerals are called **ores**. To obtain useful minerals, ores must be mined, refined, and processed. If ores are deep in the ground, they can be obtained through tunnels. If they are close to the surface, they can be obtained from open pits.

Table 7-1. Uses of Common Minerals

Group	Minerals/ Products	Common Uses
Native elements	Gold	Jewelry, coins, dental fillings
	Copper	Electrical wiring, plumbing
	Graphite	Lubricants, pencils
Mineral compounds	Hematite (iron ore)	Construction, vehicles, machinery
	Halite (rock salt)	Food, chemicals, melting ice
	Garnet	Abrasives, jewelry
	Feldspar	Porcelain, glass, ceramics
Fossil fuels	Coal	Heating, power plants, synthetics
	Petroleum	Fuels, heating, medicines, plastics

Minerals are used for a variety of applications. Iron, for example, is used to produce steel, which is then used to make such products as cars, airplanes, machinery, buildings, and even eating utensils. Aluminum is used to make cans and cookware. Copper is used in plumbing and electrical wiring. Gold and silver are used in jewelry. The amount of ore currently available in known deposits of mineral ores is called the mineral reserve. If the present rates of usage continue, some mineral reserves may be completely used up within the next 100 years.

Fossil fuels are another nonrenewable resource. Fossil fuels, such as coal, petroleum, and natural gas, formed during millions of years from the remains of dead plants and animals. Ancient plants and animals were buried under sediments when they died. Over time, heat and pressure changed the sediments into rocks and the plant and animal remains into fossil fuels.

Fossil fuels are the world's major sources of energy. Energy is the ability to do work. When fossil fuels are burned, they release energy that can be used to run machinery, automobiles, or generators that produce electricity. At present, fossil fuels are

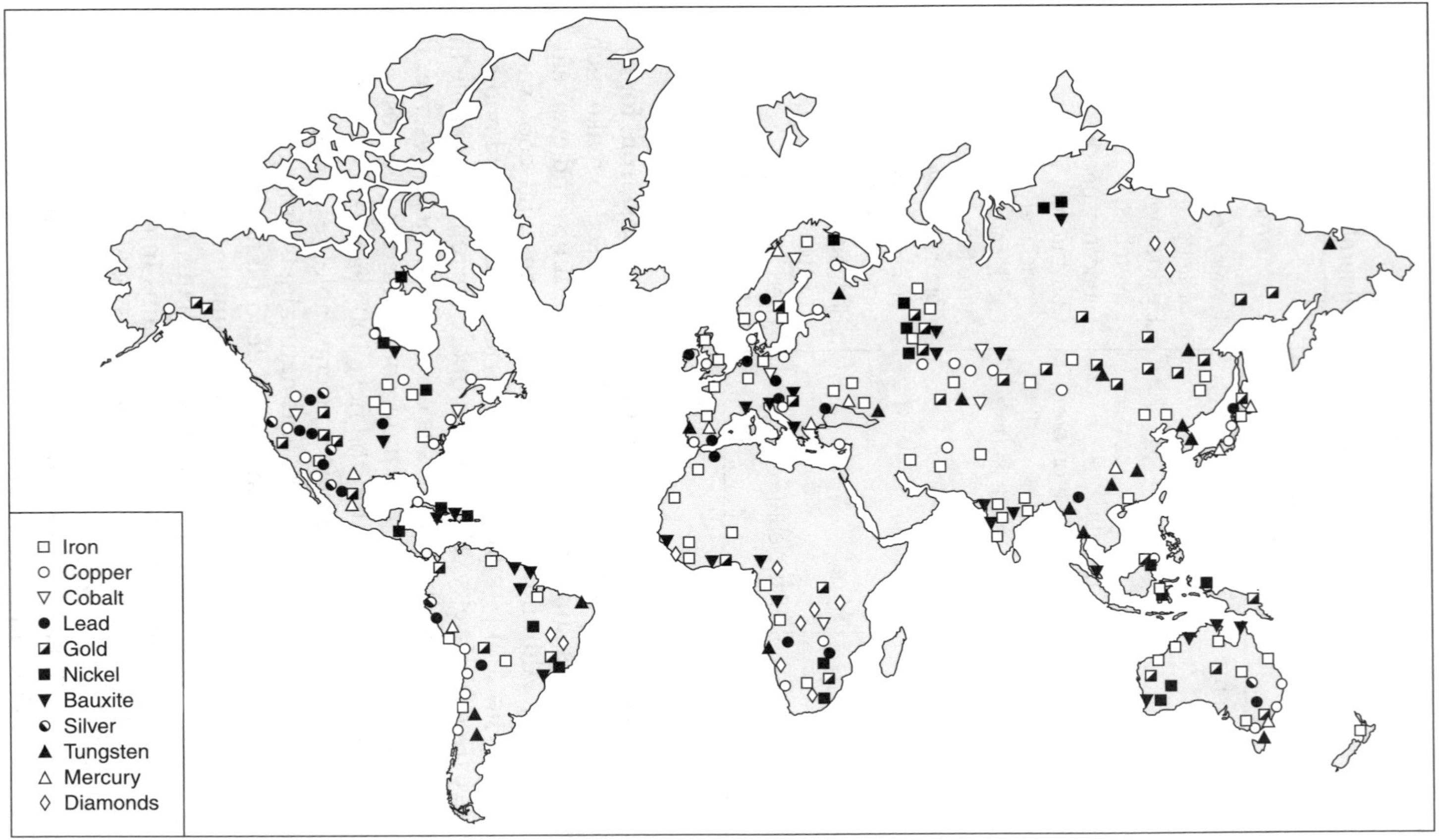

Figure 7-1. Major mineral reserves around the world.

Legend:
- □ Iron
- ○ Copper
- ▽ Cobalt
- ● Lead
- ◨ Gold
- ■ Nickel
- ▶ Bauxite
- ◐ Silver
- ◀ Tungsten
- △ Mercury
- ◇ Diamonds

being used so quickly that some scientists estimate that in fewer than 500 years, humans will have used up an amount of fossil fuels that took more than 500 million years to form.

RENEWABLE RESOURCES

Before fossil fuels, people used wood as the major source of energy. Like fossil fuels, wood can be burned to release energy. That energy can be used to perform such tasks as cooking food, warming a home, or boiling water. Wood is a renewable resource because new trees can be planted to replace those that are cut down.

Topsoil is another renewable resource. Topsoil is a thin layer of rich dirt and humus (dead and decaying organic matter) in which plants grow best. Topsoil is necessary for agricultural crops as well as forests and other habitats. New topsoil is forming constantly, but the process is so slow that it may take up to 100 years for just a few centimeters of topsoil to develop.

Water is a renewable resource. Earth's supply of water is constantly renewed, or recycled. In the **water cycle**, water moves from Earth's surface to the atmosphere and back to the surface. The water cycle involves three basic parts. In one part, the sun heats water on Earth's surface until it evaporates (changes from a liquid to a gas). Once it rises into the atmosphere, the gas (water vapor) condenses into water droplets that make up clouds. Eventually, the water droplets in the clouds fall back to Earth as rain, snow, sleet, or hail. When water falls to Earth in these forms, it is known as precipitation. After precipitation falls, it infiltrates the soil or runs into streams, lakes, and the oceans.

In some ways, air can be considered a renewable resource. Air is a mixture of several gases, but primarily nitrogen and oxygen. Air also contains carbon dioxide, which is given off during respiration (the process in which most organisms release the energy in food). Plants use the carbon dioxide and give off oxygen when they make food in the process of photosynthesis. Respiration and photosynthesis together maintain oxygen and carbon dioxide levels in the atmosphere.

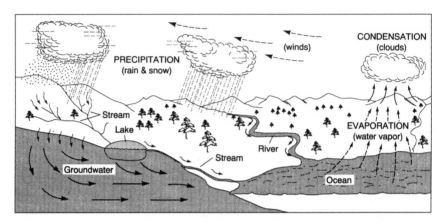

Figure 7-2. Water is continuously replaced on Earth as a result of the water cycle.

CONSERVATION OF EARTH'S RESOURCES

Conservation is the wise and careful use, protection, and preservation of Earth's natural resources, both renewable and nonrenewable.

Why do we need to protect renewable resources if Earth replaces them naturally? The reason is that even though renew-

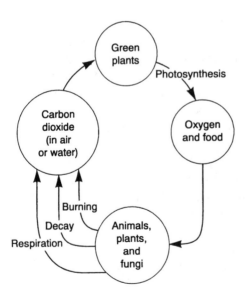

Figure 7-3. The levels of carbon dioxide and oxygen in the atmosphere are maintained by the carbon dioxide–oxygen cycle.

able resources are replaced over time, human activities alter (and sometimes reverse) Earth's natural processes. If, for example, people cut down all of the trees in an area without planting new trees, the supply of trees will diminish. Large areas of forests can be lost to human carelessness in the form of forest fires or in clearing land for agriculture.

Topsoil can be lost to erosion. Erosion is the gradual wearing away of Earth's crust. Every year, millions of acres of farmland are subject to erosion by wind and water. Poor farming techniques are a major factor in erosion. Techniques such as planting different crops on the same piece of land each year (crop rotation) and fertilization are two ways of conserving topsoil. Other threats to topsoil include the expansion of commercial development, including new buildings, roads, and industrial sites. Sometimes the construction of new dams also causes a loss of topsoil when productive agricultural lands are flooded.

We are not in danger of running out of water, but water must be conserved because the water supply is limited. Unlike trees, which can be replaced by new plants, new water is not being created. Nor can we speed the pace at which water circulates through the water cycle. We can, however, protect the natural process of the water cycle. A watershed is the area drained by a river or stream in a particular region. The soil, trees, grasses, and other plants in a watershed act together to conserve water. If the trees in the watershed are cut down or the plants are lost due to overgrazing, poor farming practices, or fire, water will flow away over the ground's surface rather than being held in the watershed. A poorly managed watershed can cause streams, springs, and wells to dry up as underground reservoirs

Figure 7-4. Farming techniques, such as terracing (left) and planting crops in strips (right), help conserve soil.

go dry. Rapid runoff can also lead to floods. In either case, the natural balance of water is upset.

Another danger to the water supply is pollution. **Pollution** occurs when part of the environment becomes unfit for use by living things. Water is polluted by pesticides, sewage, and oil spills. Substances such as mercury, lead, and other wastes also pollute water when they are released from household and industrial processes. We can protect the water supply by keeping it from becoming polluted, for example, by treating sewage and preventing oil spills.

Like water, air is often polluted by human activities, such as the burning of fossil fuels in power plants and automobile engines. The exhaust from these sources contains gases that produce smog and can lead to the gradual warming of the atmosphere. Natural sources, such as volcanic eruptions and forest fires, can add to the problem. Since these processes add carbon dioxide to the air, they upset the natural balance preserved by photosynthesis and respiration. In a similar way, the balance can be upset when rain forests are cut down, decreasing the number of plants available to produce oxygen and remove carbon dioxide from the air.

Nonrenewable resources can be conserved by extending the existing supplies. One way to conserve nonrenewable resources is to limit our use of them. We can conserve fossil fuels through

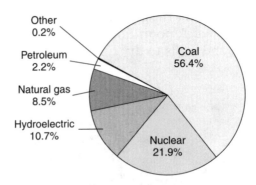

Figure 7-5. This pie chart shows which energy resources are used most commonly.

simple lifestyle changes, such as using mass transportation and smaller vehicles, insulating buildings more efficiently, and using energy-efficient lightbulbs. Other ways to conserve are to turn off appliances and lights when they are not in use, take short showers and limit running faucets, make sure appliances are operating properly, and set thermostats appropriately.

Nonrenewable resources must also be protected from abuse. Coal, for example, was once mined as if it would last forever. Early mining techniques led to great waste and environmental damage. Similarly, natural gas was once allowed to escape into the air because people did not know how to use it or did not want to incur the costs of transporting it.

Another way to conserve nonrenewable resources is to make alternative energy sources practical and cost-efficient. Solar energy is a renewable resource that does not release wastes into the environment. However, it can be expensive and it depends on location, season, and weather. Nuclear energy is another option. A tremendous amount of energy is stored within the atoms that make up matter. When the center, or nucleus, of an atom is split, that energy is released and can be used to generate electricity. Problems with this energy source include reactor accidents that release harmful radiation and wastes are dangerous for many thousands of years. Other alternative energy sources are water, wind, and tides. These sources are renewable but have their drawbacks in terms of expense and availability. So far these sources make up a very small percentage of those used most commonly throughout the world. Research into these alternative energy sources may provide viable replacements for fossil fuels in the future.

Nonrenewable resources can also be extended if they are **recycled**. When a material is recycled, it is saved, reprocessed, and used again. Through recycling, the existing supply of minerals, particularly the metals, will last longer. If, for example, copper from buildings and vehicles that are to be destroyed is recycled, less copper would need to be mined. Similarly, aluminum, glass, iron, and plastics can be reused if we collect and separate them out of our trash for recycling.

The key to protecting Earth's natural resources for the future is a committed and consistent conservation effort.

QUESTIONS

Multiple Choice

1. An example of a nonrenewable resource is *a.* wood
   b. petroleum   *c.* soil   *d.* water.
2. The most commonly used energy resource is   *a.* solar
   energy   *b.* energy from coal   *c.* nuclear energy   *d.* tidal
   energy.
3. An example of a renewable resource is   *a.* coal   *b.* natural
   gas   *c.* iron ore   *d.* fresh water.
4. During photosynthesis, plants give off   *a.* oxygen
   b. nitrogen   *c.* carbon dioxide   *d.* minerals.
5. Recycling would most help preserve our supplies of
   a. coal   *b.* petroleum   *c.* oxygen   *d.* minerals.

Fill In

6. _____ resources are used at a faster rate than the rate at
   which they are produced in the environment.
7. Coal and petroleum are _____, which were formed from
   the remains of plants and animals millions of years ago.
8. _____ resources can be replaced by the environment.
9. _____ is the wise use and protection of natural resources.
10. Minerals can be _____, or used again, to preserve their
    supplies.

Free Response

11. What is a natural resource?
12. How are renewable and nonrenewable resources alike and
    how are they different?
13. What are some of the uses of minerals? How can minerals
    be conserved?
14. What are fossil fuels? How are they used? How might
    fossil fuels be conserved?
15. Why are air and water considered renewable resources?
    How can their supplies be decreased?

Portfolio

I. Prepare a report on a particular mineral resource, telling where it is found, how it is processed, and how we use it.

II. Investigate ways we can meet our needs for a particular resource as supplies of this material become more scarce.

III. Suggest new laws that would help to preserve resources without a major impact on our economy or our comfort.

IV. Investigate recycling in your community.

V. Prepare a report about a resource that is presently in great demand and is in danger of being used up. Speculate on how we can overcome this problem.

Chapter 8
Plate Tectonics

So far you have been reading about measuring Earth, mapping Earth, and the importance of Earth's resources. What you may have realized by now is that Earth is a dynamic planet that is constantly changing. Throughout its long history, Earth's surface has been twisted, lifted, bent, and broken. Forces beneath the surface are constantly changing Earth's appearance. In this chapter, you will learn about the structure of Earth and why the surface changes. You will be exploring Earth like a **geologist**, a scientist who studies the forces that make and shape Earth.

EARTH'S INTERIOR

Earth can be divided in three main layers: the crust, the mantle, and the core. The outer layer of Earth is the **crust**. The average thickness of the crust ranges from 5 to 30 kilometers. The crust beneath the ocean is called oceanic crust. It is made up primarily of basalt, which is a dark, dense rock. The crust that forms Earth's large landmass, or continents, is called continental crust. It consists primarily of granitic rocks. The crust is thinnest beneath the ocean and thickest under high mountains.

Beneath the crust is a layer of hot rock called the **mantle**. The entire mantle is about 2900 kilometers thick. The uppermost part of the mantle is very similar in composition to the crust. This upper layer of the mantle and the crust form a layer known as the **lithosphere**. Beneath the lithosphere is a softer layer of rock known as the **asthenosphere**. High heat and pres-

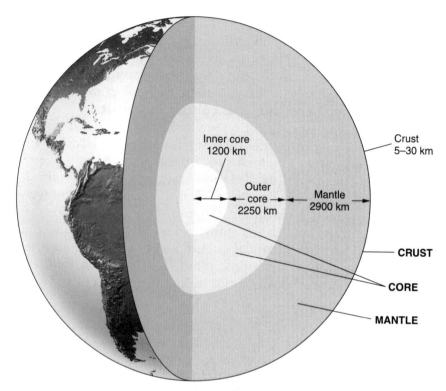

Figure 8-1. Layers of Earth.

sure make rock in the asthenosphere soft enough to flow slowly. The lithosphere actually floats on top of the asthenosphere.

Below the asthenosphere, solid mantle extends to Earth's core. The **core** is the innermost layer of Earth. The core consists of two parts—the outer core and the inner core. The outer core is a layer of molten metal, which behaves like a thick liquid. The inner core is dense and solid.

CONTINENTAL DRIFT

If you look on a map, you might notice that some of the continents look as if they might fit together like the pieces of a puzzle. When the first accurate maps were created, many people made the same observation and wondered why this was the case. Early in the twentieth century, a German scientist named

Alfred Wegener suggested that the continents had once been joined together as a single landmass, which he named Pangaea (meaning all-Earth). He proposed that Pangaea broke apart forming the present continents. Over tens of millions of years, the continents slowly drifted to their present positions. This theory has come to be known as **continental drift**.

This idea was so radical that few geologists took it seriously. But Wegener supported his theory with several pieces of evidence. First, he used landforms to support his theory. Wegener showed that the shoreline of South America would fit remarkably well with the shoreline of Africa if the Atlantic Ocean were closed up. In fact, a mountain range in South Africa lines up with a mountain range in Argentina. In addition, Brazilian coal fields match up with identical coal fields in Africa. Second, he pointed to a fossil of a fernlike plant that has been found in Africa, South America, Australia, India, and Antarctica. Third,

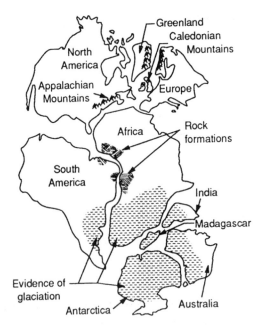

Figure 8-2. When the continents are drawn close together, land features and shapes indicate that the continents were once joined.

Wegener noted evidence that certain continents had once been exposed to different climates. For example, deep scratches in rocks showed that glaciers (huge sheets of ice) once covered South Africa. The climate in South Africa today is much too mild for glaciers to form. Rather than suggesting that the climate had changed, Wegener proposed that South Africa had once been much closer to the South Pole.

SEAFLOOR SPREADING

Although a few geologists accepted Wegener's theory, most scientists found it too absurd to think of the continents plowing through the solid rock of the ocean floor. Further evidence had to come from the ocean floor itself. Extended exploration of the ocean depths, made possible by technological advances achieved during World War II, revealed a great system of underwater mountains that encircle Earth like the seams on a baseball. This underwater mountain range is known as the system of **mid-ocean ridges**. It is a continuous mountain chain that extends more than 64,000 kilometers. Most of the mountains in the mid-ocean ridge are under water. In a few places, however, the mountains rise up through the ocean surface. The island of Iceland, for example, is part of the mid-Atlantic ridge.

In 1960, an American geologist named Harry Hess reconsidered Wegener's theory in light of his research into the mid-ocean ridge. He suggested that molten material from the mantle rises and erupts through a valley that runs along the center of the mid-ocean ridge. The molten material then spreads out, pushing older rock to both sides. As the molten material cools, it forms a strip of solid rock in the center of the ridge. Over time, more molten material erupts. This material splits apart the strip of solid rock that formed before, pushing it to either side. Hess called this process **seafloor spreading**.

Several other forms of evidence supported Hess's theory of sea-floor spreading. First, scientists found evidence that new material is indeed erupting along the mid-ocean ridge. Scientists found strange rocks shaped like pillows, which can form only when molten material hardens quickly after erupting under water.

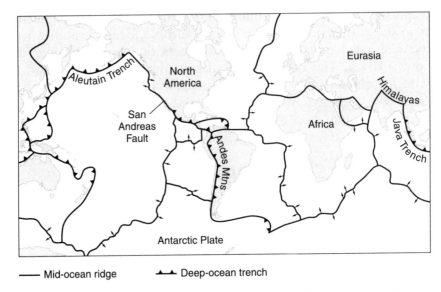

— Mid-ocean ridge   ▲▲ Deep-ocean trench

Figure 8-3. The mid-ocean ridge on Earth's ocean floor. Arrows show the relative direction of plate motion.

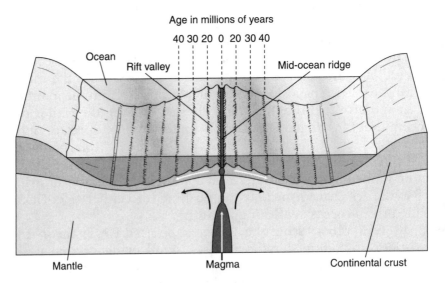

Figure 8-4. As oceanic crust spreads away from the ridge, new crust is formed.

Now recall from Chapter 3 that Earth acts like a huge magnet. The rock of the ocean floor contains iron, which is a magnetic material. When the rock is still soft, the iron bits line up in the direction of Earth's magnetic field. As the rock hardens, the iron bits are locked in place like a magnetic memory. The direction of the magnetism depends on the location of Earth's magnetic poles. Earth's magnetic poles have reversed themselves several times, every 300,000 years or so. In other words, the north magnetic pole became the south magnetic pole and the south magnetic pole became the north magnetic pole. If the entire ocean floor was formed at the same time, all of the rock should be magnetized in the same direction. But this is not the

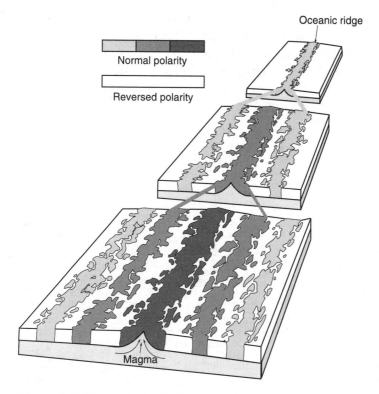

Figure 8-5. Magnetic "zebra" stripes form as a result of reversals in the direction of Earth's magnetic field.

case. Instead, scientists found stripes. The magnetic material in one stripe of rock points in one direction, while a parallel stripe points in the other direction. These stripes indicate that the rocks could not have been formed at the same time.

The final evidence came from rock samples obtained by drilling into the ocean floor. When scientists determined the age of the rocks in the samples, they found that farther away from the ridge, on either side, the rocks became older. The youngest rocks are always found near the center of the ridges.

How can the ocean floor keep spreading if Earth is not getting any larger? As new floor is created, some ocean floor plunges into deep underwater canyons called **ocean trenches**. In these trenches, subduction takes place. **Subduction** is the process in which the ocean floor sinks back into the mantle and melts. The processes of seafloor spreading and subduction work together much like a conveyer belt. New rock forms at the mid-ocean ridge, moves across the ocean, and sinks into a deep-ocean trench. Unlike oceanic crust, which is drawn back into Earth during subduction, the continental crust, which is less dense than oceanic crust, does not sink. Instead, continental rocks crumple, forming folds, faults, and mountains.

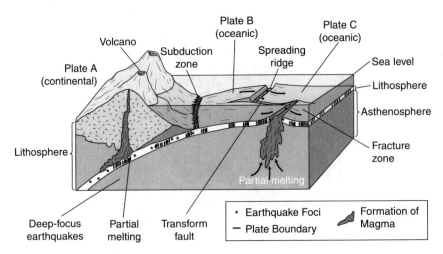

Figure 8-6. As new ocean floor is created at the mid-ocean ridge, older ocean floor sinks into a trench where it blends back into the mantle.

PLATE TECTONICS

Where and why does the mid-ocean ridge form? Earth's crust is not an unbroken layer. Instead, it has many cracks in it, much like a cracked eggshell. In 1965, a Canadian scientist named J. Tuzo Wilson suggested that the lithosphere is broken into separate sections called lithospheric **plates**. There are about a dozen major plates and a number of small plates. These plates are approximately 100 kilometers thick, which is very thin when compared with the 6000-kilometer radius of Earth. Most major plates include both continental crust and oceanic crust.

Wilson proposed a theory that combines continental drift, seafloor spreading, and lithospheric plates into a single theory known as **plate tectonics**. According to this theory, lithospheric plates move, carrying the continents or parts of the ocean floor along with them. Recall that the lithosphere floats on the denser

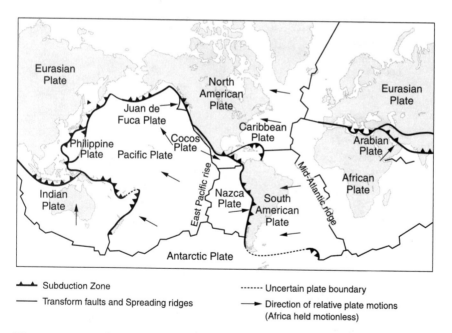

Figure 8-7. Earth's crust is made up of numerous lithospheric plates, which move in various directions.

asthenosphere. Currents, known as convection cells, in the asthenosphere are believed to cause the plates to move.

The core is hotter than the crust. Heat from the core causes heated rock to rise in the asthenosphere, usually at the mid-ocean ridges, and move along the boundary of the lithosphere and asthenosphere. This rock cools slightly as it moves along the boundary and sinks toward the core, usually at subduction zones. This slow movement causes plates to move.

As the plates move, they collide, pull apart, or slide past each other. The edges of different plates meet at plate boundaries. There are three different kinds of plate boundaries: transform boundaries, convergent boundaries, and divergent boundaries.

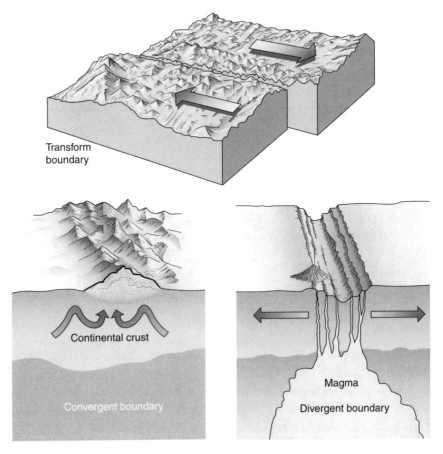

Figure 8-8. Three types of plate boundaries.

A **transform boundary**, or sliding boundary, is a place where two plates slip past each other. Transform boundaries are often associated with earthquakes. A **convergent boundary** occurs where two plates come together, or converge. Convergent boundaries often result in the formation of mountains. A **divergent boundary** occurs where two plates diverge, or move apart. Most divergent boundaries occur at the mid-ocean ridges.

You may wonder why you do not see the continents moving. The reason is that they move very slowly. In fact, until recently scientists couldn't measure relative changes in the positions of the continents because they were unable to detect changes of a few centimeters over a distance of thousands of kilometers. However, satellite technology has allowed scientists to confirm this motion by direct measurements. Some plates move as fast as 10 centimeters per year, although speeds of 2 to 4 centimeters per year are more common. Over millions of years, this movement carries the continents thousands of miles.

PARADIGM SHIFTS

It is important to recognize that our understanding of the world can change as we develop new ways to explore and analyze our observations. In about a decade, the theory of plate tectonics went from the eccentric idea held by a few earth scientists to a major shift in the thinking of most geologists. This constituted a major paradigm shift. A **paradigm** is a coherent set of principles and understandings. A paradigm shift occurs when new observations or revised understandings lead us away from our traditional way of thinking about nature and into a new set of principles.

Many paradigm shifts have occurred over time. For example, about 200 years ago scientists working in the Alps of Europe observed uniquely shaped hills and bedrock features exposed as the local glaciers melted back. They realized that they had observed similar features far from the Alps and its present glaciers. A great deal of debate took place among geologists before they could accept the idea that glaciers recently covered most of Europe and the northern part of the United States. However, the new theory explained a wide range of geologic features that

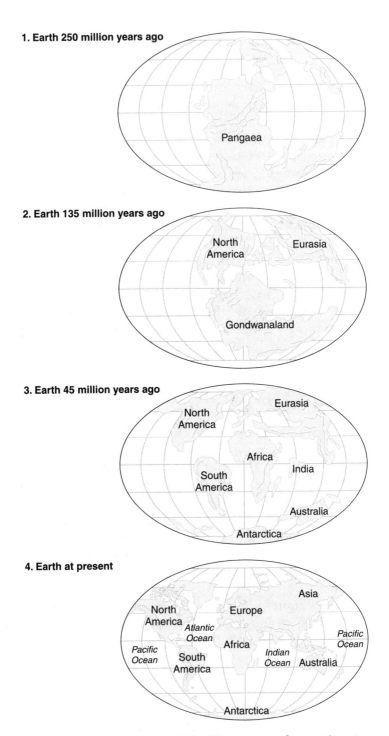

1. Earth 250 million years ago

Pangaea

2. Earth 135 million years ago

North America

Eurasia

Gondwanaland

3. Earth 45 million years ago

Eurasia

North America

Africa

India

South America

Australia

Antarctica

4. Earth at present

Asia

North America

Europe

Atlantic Ocean

Pacific Ocean

Pacific Ocean

Africa

South America

Indian Ocean

Australia

Antarctica

Figure 8-9. Over the past 200 million years, the continents have moved to their present locations.

could not be understood in terms of the previously accepted ideas about geologic history.

Earlier paradigm shifts occurred when people accepted the fact that the sun is the center of the solar system and that Earth is round instead of flat. There are many such examples in history, and there will probably be many more in the future. So always feel free to think and question.

QUESTIONS

Multiple Choice

1. The crust and the upper portion of the mantle make up the   *a.* asthenosphere   *b.* inner core   *c.* outer core   *d.* lithosphere.

2. The mid-ocean ridge rises above the water to form   *a.* Hawaii   *b.* Alaska   *c.* Iceland   *d.* Bermuda.

3. As the distance from the mid-ocean ridge increases, the age of the rock   *a.* increases   *b.* decreases   *c.* remains the same   *d.* is unpredictable.

4. The continental crust is not drawn into the mantle by subduction because compared to oceanic crust it is   *a.* hotter   *b.* more dense   *c.* less dense   *d.* larger.

5. A transform boundary occurs where two lithospheric plates   *a.* are moving apart   *b.* are moving toward each other   *c.* are sliding past each other   *d.* have come to a standstill.

Fill In

6. Scientists who study the forces that shape Earth are called _____.

7. The _____, which is at the center of Earth, is divided into an inner and an outer section.

8. The sea floor spreads along underwater mountain ranges, known as the _____.

9. During subduction, part of the ocean floor plunges into a (an) _____.

10. A set of principles or understandings is known as a (an) _____.

Free Response

11. Describe the major layers of Earth. Include a diagram in your response.

12. Summarize Wegener's theory of continental drift. What evidence did Wegener use to support his theory?

13. Explain how the sea floor can spread at mid-ocean ridges. Cite evidence to support this theory.

14. Identify the three types of plate boundaries and describe how they might cause changes to Earth's surface.

15. What is a paradigm shift? Why might it be important to recognize that such shifts can occur?

Portfolio

I. Prepare a report about the life of Alfred Wegener and how he developed his ideas about continental drift.

II. Prepare a poster listing the various pieces of evidence that support the theory of plate tectonics. For each item in your list, briefly explain how it supports this theory and where it can be observed.

III. Investigate evidence in your own region that supports the motion of the continents.

Chapter 9

Earthquakes

EARLY ON THE MORNING of March 27, 1964, the residents of Anchorage, Alaska, felt the gentle rolling motion that marked the onset of one of the most violent events in modern history. For five minutes the ground shook and heaved as one side of the city's main shopping street dropped 3 meters. By the time it was over, a housing development had slid into the bay, streets were moved, buildings collapsed, and huge cracks opened in the ground. Anchorage had suffered a devastating earthquake. In this chapter, you will learn about earthquakes, why they occur, how they are measured, and the damage they cause.

WHAT IS AN EARTHQUAKE?

An **earthquake** is the shaking of Earth's crust due to a sudden release of energy. Earthquakes are caused by stress that builds up between two lithospheric plates. Stress is a force that acts on rock. When two plates slide next to each other, friction usually prevents the plates from moving. Friction is a force that opposes motion. Instead of moving, the stresses on the plates cause the plates to change shape. Eventually, the stresses become great enough to overcome friction, and the plates suddenly move. This movement causes an earthquake. The earthquake releases the stress, however, stress will probably build again over time.

Earthquakes generally occur along faults. A **fault** is a break or crack in Earth's crust along which movement has occurred. In

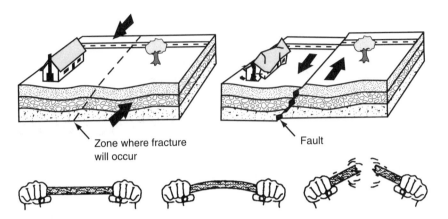

Figure 9-1. An earthquake releases stress that has built up along a fault.

California, two plates are sliding past each other along the San Andreas Fault. The point on the fault plane where movement first occurs is called the **focus**. The point on the surface directly above the focus is the **epicenter**. Earthquakes are strongest near the epicenter.

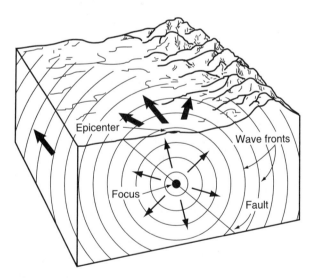

Figure 9-2. An earthquake occurs when motion begins at the focus. The epicenter is located directly above the focus.

SEISMIC WAVES

An earthquake produces waves in Earth much like a stone tossed into a pond creates waves in water. Earthquake waves are called **seismic waves**. Like all waves, seismic waves transmit energy. There are three basic types of seismic waves: primary, secondary, and surface waves. **Primary waves**, or P-waves, alternately squeeze and stretch the materials through which they pass. This type of wave is known as a longitudinal wave. It is much like the spring. P-waves travel through any material—solid, liquid, or gas. **Secondary waves**, or S-waves, cause particles to move at right angles to the direction in which the waves are traveling. This type of wave is known as a transverse wave. It is much like the wave on a rope. S-waves travel through solids, but not through liquids and gases. When P- and S-waves reach Earth's surface, they form **surface waves.** Surface waves cause rock to move in a circular fashion.

In terms of speed, P-waves travel about twice as fast as S-waves. Surface waves travel more slowly than either P- or S-waves. The difference in the speeds of P-waves and S-waves gives seismologists a method of determining the distance to the epicenter of an earthquake. (A **seismologist** is a person who studies earthquakes.) P-waves always arrive at a given point before S-waves. The difference in arrival times is found by simply subtracting the arrival time for P-waves from the arrival time for S-waves. The farther the reference point is from the epicenter, the greater the difference in arrival times between P- and S-waves. The relationship between P and S arrival times and the distance to the epicenter is shown on a time-travel graph. A seismologist finds the place on the graph where the lines for P- and S-waves are separated by the time difference he or she measured. By reading the distance from the graph, the seismologist can find the distance to the epicenter.

The method gives the distance to the epicenter but not the exact location. Imagine, for example, that a friend tells you she is 10 kilometers away. She could be 10 kilometers in any direction. If you drew a circle with a radius of 10 kilometers around yourself, she could be anywhere on the circle. In a similar way, a seismologist at one recording station can limit the location of

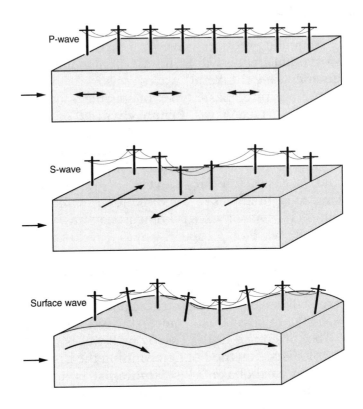

Figure 9-3. A P-wave causes particles to vibrate in the same direction in which the wave is moving. An S-wave causes particles to move side to side. A surface wave causes a circular motion on Earth's surface. The waves are moving from left to right.

the epicenter to a circle around the station. The radius of the circle is the distance to the epicenter. By adding the data from two additional recording stations, seismologists can determine the exact location of the epicenter.

Seismic waves are detected and recorded by a device called a **seismograph**. A seismograph detects ground movements—either horizontal or vertical motions. A mechanical seismograph consists of a heavy weight suspended from a base and a drum that is turned slowly. A pen attached to the weight rests on paper wrapped around the drum. When an earthquake occurs, seismic

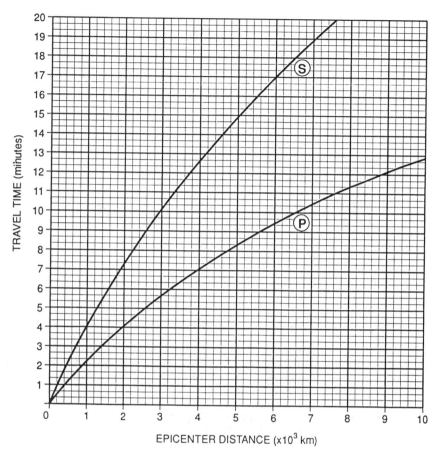

Figure 9-4. Time-travel graph for P- and S-waves.

waves shake the ground beneath the drum. The pen does not shake because it is attached to the heavy weight. The shaking drum causes a jagged line to be drawn on the paper. The height of the jagged lines is related to the energy of the earthquake. The stronger the earthquake, the larger the up-and-down tracing. The set of jagged lines recorded on paper is called a **seismogram**. There are different types of seismographs, depending on the types of earthquakes they are intended to record. Most modern seismographs are electronic devices rather than mechanical, which means they convert ground movement into electronic signals.

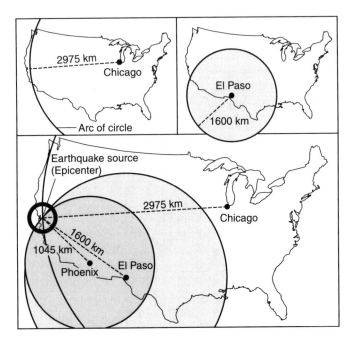

Figure 9-5. When the distance to the epicenter of an earthquake is determined by three recording stations, the exact location of the epicenter can be found.

MEASURING EARTHQUAKES

There are two kinds of measurements generally applied to earthquakes: intensity and magnitude. The **Mercalli scale** is a measure of intensity. An earthquake's intensity is the strength of ground motion in a given location. The Mercalli scale is not a precise measurement, but rather a description of how the earthquake affected people, buildings, and land. The Mercalli scale extends from intensity I to XII.

The **Richter scale,** developed in 1935, is a rating of the magnitude of seismic waves as measured by a particular type of seismograph. In recent years, another method of determining magnitude has been developed. The **seismic moment** estimates the total energy released by an earthquake. Both of these scales are logarithmic. That is, each change of 1 on the magnitude scale is a change by a factor of 10 in strength. So a magnitude 4 earthquake causes ten times as much shaking as a magnitude 3

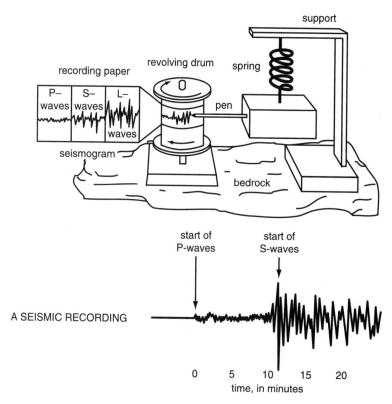

Figure 9-6. Traditional seismographs record shaking of the ground with a large mass suspended on a spring. The seismic recording is called a seismogram.

Table 9-1. Mercalli Scale

Intensity	Effects
I–II	Almost unnoticeable.
III–IV	Vibrations are noticeable and unstable objects are disturbed.
V–VI	Dishes can rattle. Books may be knocked off of shelves. Slight damage.
VII–VIII	Shaking is obvious, often prompting people to run outside. Moderate to heavy damage.
IX–X	Buildings knocked off foundations. Cracks formed in ground. Landslides may occur.
XI–XII	Wide cracks appear in ground. Waves seen on ground surface. Severe damage.

earthquake, or a tenth the shaking of a magnitude 5. Although magnitude scales have no upper limit, there does seem to be a limit to the stress that rocks can hold before they break. For this reason, it seems unlikely that we will ever observe tremors of 10 or more on magnitude scales.

EARTHQUAKE DAMAGE

Seismologists have a saying, "Earthquakes don't kill people, buildings do!" The greatest toll in human life generally occurs where the density of population is high and people live in unreinforced buildings. The shaking caused by an earthquake can move buildings up and down as well as side to side. Most buildings collapse as a result of the side-to-side movements related to S-waves. In addition, ground shaking can cause the soil under a building to settle or even become liquefied, or muddy. When this happens, the building is no longer supported and collapses. Earthquakes can also cause landslides, avalanches, and giant waves called tsunamis, which are generated by underwater landslides.

EARTHQUAKE PREDICTION

Where is the next earthquake likely to occur? Unfortunately, this question is not easy to answer. Most people think of California when they think of earthquakes. What they do not realize is that many areas are at just as much risk as California. Figure 9-7 shows areas of risk in the United States. Although earthquake risk is not the same everywhere, very few locations are without some risk.

With so many people living in areas of earthquake risk, there is a tremendous need for an effective method of predicting earthquakes. Predicting an earthquake involves determining where and when the earthquake will occur, as well as how strong it will be. Seismologists have looked at a great variety of factors to serve as warnings of large earthquakes. They have monitored the height of the water levels in wells, small changes in the tilt of the land near major faults, and even the behavior of animals. In re-

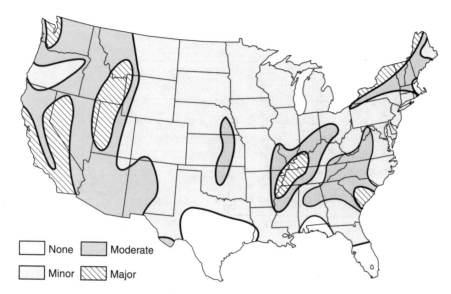

Figure 9-7. Earthquake hazard is based on historical records. Although earthquakes are less common in the central and eastern United States than they are along the Pacific Coast, there have been major earthquakes in all three regions.

cent years there has been enthusiasm about recording minor earthquakes that sometimes come before a major tremor. At this time, none of these factors seems to be a consistent predictor of a major earthquake.

LEARNING FROM EARTHQUAKES

In Chapter 8 you learned about the inner layers of Earth. Yet geologists could not learn this information by digging a 6000-kilometer-deep hole to the center of Earth. Instead, geologists have to rely on indirect methods of observation. The most significant method of indirectly studying Earth's interior is by analyzing seismic waves.

Scientists can use seismograms along with time-travel graphs to determine if seismic waves have changed speed or direction. A change in the speed of a wave indicates that the wave is traveling through a different substance, or medium. Waves travel faster through solids than they do in liquids. For example,

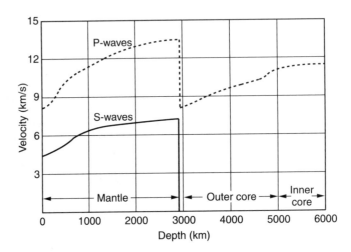

Figure 9-8. Analysis of seismic waves has provided scientists with information about Earth's mantle and core.

look at the graph comparing wave velocity to depth below Earth's surface. At about 2900 kilometers, both types of waves suddenly slow down. The S-waves stop moving altogether. What might this indicate? Since S-waves do not travel through liquids, the material directly below 2900 kilometers must be liquid. This is how scientists determined that Earth's outer core is liquid. Now notice what happens to the P-waves as they continue to travel through the core. The P-waves speed up again. Since waves travel faster through solids than liquids, this indicates that the inner core is solid.

QUESTIONS

Multiple Choice

1. The point on the fault at which movement first occurs is the   *a.* epicenter   *b.* focus   *c.* plate   *d.* seismogram.
2. The fastest seismic waves are   *a.* S-waves   *b.* L-waves   *c.* P-waves   *d.* E-waves.
3. The Mercalli scale measures   *a.* magnitude   *b.* intensity   *c.* distance   *d.* location.

4. When compared with a reading of 5 on the Richter scale, an earthquake of magnitude 6 is   *a.* 25 percent stronger *b.* twice as strong   *c.* four times as strong   *d.* ten times as strong.

5. Which type of seismic wave causes the most damage to buildings?   *a.* primary waves   *b.* secondary waves   *c.* surface waves   *d.* water waves

6. How many recording stations are necessary to find the location of an earthquake's epicenter?   *a.* 1   *b.* 2   *c.* 3 *d.* 10

Fill In

7. The shaking of Earth's crust due to the movement of lithospheric plates in a(an) _____.

8. A(an) _____ is a break in Earth's crust along which movement has occurred.

9. The _____ is the point on Earth's surface directly above the focus.

10. When P- and S-waves reach the surface they form _____ waves.

11. Seismic waves are detected and recorded on an instrument known as a _____.

12. The magnitude of an earthquake can be measured on the _____ scale, which was established in 1935.

Free Response

13. How are earthquakes related to plate tectonics?

14. Differentiate among P-, S-, and surface waves.

15. How can a seismologist use a graph to determine the distance to the epicenter of an earthquake?

16. Why must data from three recording stations be used to determine the location of the epicenter of an earthquake?

17. What is the Mercalli scale?

18. What have scientists learned about Earth's interior from earthquakes?

Portfolio

I. On a very long strip of paper, make two lines 1 millimeter apart to represent the difference between an earthquake of magnitude 3, which is about the smallest that most people feel, and magnitude 4. Label these lines M = 3 and M = 4. Make another mark ten times as far from M = 4 (1 cm) and label it M = 5. Continue up the scale, each interval ten times the previous one, until you run out of paper.

II. Prepare a report about a major earthquake. Base your report on newspaper and scientific sources.

Chapter 10

Volcanoes

ONE OF THE MOST SPECTACULAR and violent events on Earth is the eruption of a volcano. Some volcanic eruptions, such as Vesuvius in A.D. 79 and Krakatau in 1883, are known for destroying entire towns and the people in them. Others are known for creating beautiful landforms such as the Hawaiian Islands. In this chapter, you will learn about volcanoes and how they are formed.

WHAT IS A VOLCANO?

A **volcano** is an opening in Earth's surface through which magma erupts. Recall that magma is molten rock underground. Magma is formed when rock melts as a result of high temperatures and pressure. Magma is found in the asthenosphere of Earth's mantle. Magma can also form at the boundaries of lithospheric plates. Once magma forms, it flows upward through cracks in the rock above it. Magma rises until it becomes trapped under layers of rock or reaches the surface. If it reaches the surface, it erupts through a volcano. Magma that erupts through a volcano is called lava. Recall that when lava cools, it forms igneous rock.

Figure 10-1 shows the outline of a volcano. You can see that below the volcano magma collects in a pocket called a magma chamber. The magma moves through a long break in the ground that connects the magma chamber to Earth's surface. Magma

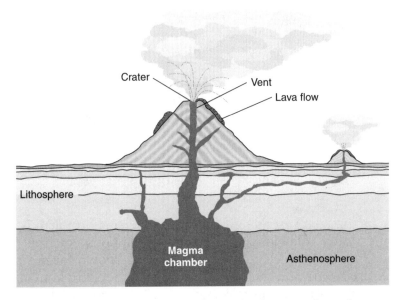

Figure 10-1. Magma from Earth's interior escapes through a volcano. The points at which magma pours onto Earth's surface as lava are called vents. (Not drawn to scale.)

escapes through an opening called a **vent**. A **crater** is a bowl-shaped area that may form at the top of a volcano around the vent.

VOLCANIC ERUPTIONS

Some volcanoes erupt quietly. Others erupt explosively. The difference depends on the type of magma involved. Magmas can be thick and flow slowly, or thin and flow easily. Quiet eruptions usually involve thin magma. Explosive eruptions usually involve thick magma that builds up until it explodes through the vent.

One factor that determines how easily the magma flows is the amount of silica it contains. Recall from Chapter 4 that silica is a material that contains oxygen and silicon. The more silica magma contains, the thicker it is. The type of eruption also depends on the amount of dissolved gases in the magma. Gases that dissolve in the magma, such as water vapor, carbon dioxide, and sulfur dioxide, are given off as the magma erupts. As the magma reaches the surface, the gases within it form bubbles much like those in a carbonated beverage. Magmas containing large amounts of dissolved gases tend to produce more explosive eruptions than magmas containing small amounts of gases.

Explosive eruptions cause the lava to break into fragments. The smallest fragments are called volcanic ash. The largest fragments, which can range from the size of a baseball to the size of a car, are called bombs and blocks.

VOLCANIC STRUCTURES

Volcanic mountains and other volcanic structures form when magma erupts and cools at Earth's surface. There are three types of volanoes: shield, cinder cone, and composite.

A **shield volcano** is a volcanic mountain with a wide base and gently sloping sides. Shield volcanoes form from repeated flows of thin lava. **Cinder cone volcanoes** are steep, cone-shaped mountains. These volcanoes form from explosive eruptions in which fragments build up in a steep pile around the vent. **Composite volcanoes** are tall, cone-shaped mountains in which layers of lava alternate with layers of ash. This happens when smooth lava flows alternate with explosive eruptions of lava fragments.

Occasionally, volcanic eruptions do not form mountains. Instead, some eruptions form high, level areas called **lava plateaus**. These are created when lava flows out of several long cracks in an area. The thin lava travels far before cooling and becoming solid. During each new eruption, lava flows on top of earlier lava. After millions of years, the layers of lava form high plateaus.

Another formation that might be created by a volcano is a caldera. A **caldera** is a hole left by the collapse of a volcanic mountain. Calderas are formed when enormous eruptions release all of the magma beneath the volcano. The mountain then becomes a hollow shell. With nothing to support it, the mountain collapses.

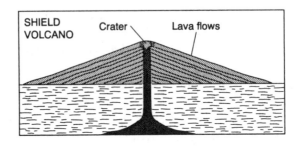

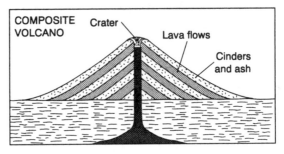

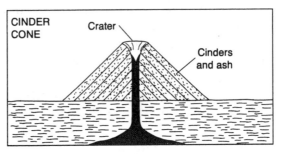

Figure 10-2. Three types of volcanoes.

WHERE DO VOLCANOES FORM?

Look at the location of volcanoes in Figure 10-3. Notice how volcanoes occur in belts along plate boundaries. One major belt, found along the rim of the Pacific Ocean, is called the Ring of Fire. Some volcanoes form at divergent plate boundaries, such as the mid-ocean ridge. Eruptions from these volcanoes are sometimes known as rift eruptions because the valley along the center of the mid-ocean ridge is called a rift valley. At these locations, lava flows through cracks in the ocean floor.

Most of the world's volcanoes occur at subduction zones. Recall that during subduction, oceanic crust sinks through deep-

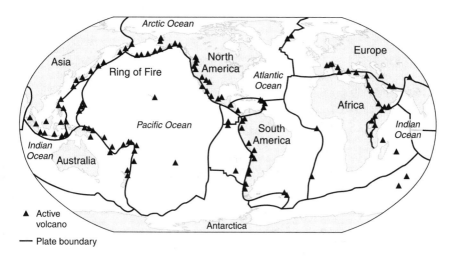

Figure 10-3. Most of Earth's active volcanoes are along plate boundaries.

ocean trenches into the mantle. The crust then melts into magma. When this magma rises to the surface, volcanoes are formed. Volcanoes such as these have formed strings of islands, such as the islands of Japan and Indonesia.

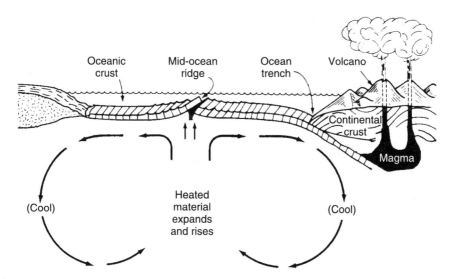

Figure 10-4. Explosive volcanoes often occur where one plate is sinking below another. As oceanic crust melts, some magma rich in gas rises through cracks to the surface.

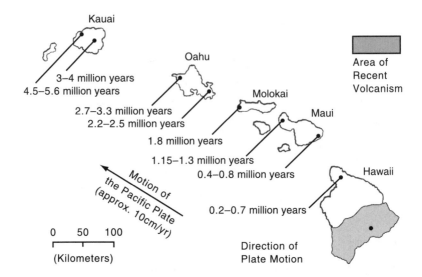

Figure 10-5. The progressive ages of volcanic activity in the Hawaiian Islands shows the slow movement of the Pacific Plate over a hot spot deep within Earth.

Not all volcanoes form at plate boundaries. Some volcanoes result from hot spots. A **hot spot** is an area where magma melts through the crust in the middle of a lithospheric plate. Hot spots remain in the same location even though the lithospheric plate above them moves. This produces a chain of extinct (no longer active) volcanoes. The Hawaiian Islands were formed as the Pacific Plate moved over a hot spot. The age of the islands increases as you move northwest from Hawaii to Kauai. This increasing age indicates plate motion.

QUESTIONS

Multiple Choice

1. Magma erupts through a volcanic opening known as a
   a. crater   *b.* vent   *c.* magma chamber   *d.* caldera.
2. The islands of Indonesia were formed by volcanoes caused
   by   *a.* a hot spot   *b.* a rift eruption   *c.* subduction
   d. an earthquake.

3. The Hawaiian Islands were formed by *a.* a hot
   spot *b.* a rift eruption *c.* subduction *d.* an
   earthquake.
4. A volcanic mountain with a wide base and gently sloping
   sides is a *a.* shield cone *b.* cinder cone *c.* composite
   cone *d.* caldera.
5. A hole left by the collapse of a volcano is a *a.* shield
   cone *b.* crater *c.* composite cone *d.* caldera.

Fill In

6. Molten rock, or _____, is formed when rock melts from
   high temperature and pressure.
7. An opening on Earth's surface through which lava erupts
   is called a(an) _____.
8. Explosive eruptions create lava fragments, the smallest of
   which is volcanic _____.
9. A volcano in the middle of a lithospheric plate is the result
   of a(an) _____.
10. A(an) _____ is a high, level area formed from repeated
    volcanic eruptions.

Free Response

11. What are two factors that determine the type of eruption
    from a volcano?
12. How are shield volcanoes different from cinder cone
    volcanoes?
13. What happens if a volcano releases the magma that sup-
    ports the surface rock?
14. How do volcanoes form at rift valleys and subduction
    zones?
15. Explain how the Hawaiian Islands formed.

Portfolio

I. On a map of the world, plot the locations of volcanoes and
   earthquakes. Explain any patterns that you discern.

   II. Construct a model of one of these types of volcanoes.

  III. Investigate and report on a famous volcanic eruption.

Chapter 11

Landscapes

Eartʜ's surface has changed greatly over billions of years. The result is an array of interesting and often beautiful landscapes that arise as Earth's surface is lifted, faulted, crumbled, and eroded. In Chapter 2 you learned that the shape of the land is called its topography. A large area of land where the topography is similar is called a **landscape** or **landform region**. A **landform** is a feature formed by the processes that shape Earth's surface, such as the movement of lithospheric plates. A landscape is described by its landforms, elevation, and relief. **Elevation** is the height above sea level. The difference in elevation between the highest and lowest parts of an area is its **relief**. In this chapter, you will learn about the three basic types of landscapes: plains, plateaus, and mountains.

PLAINS

Plains are flat land areas with low relief and low elevation. Plains are generally supported by flat-lying sediments or sedimentary rocks. If there are hills in a plains region, they are likely to be small and isolated. Plains landscapes are characterized by broad rivers and meandering streams.

A plain that lies along an ocean coast is called a coastal plain. Much of the Atlantic and Gulf coasts of the United States are bordered by coastal plains that slope gently toward the sea. A plain that is not located near a coast is called an interior plain. The broad interior plain of North America is called the Great

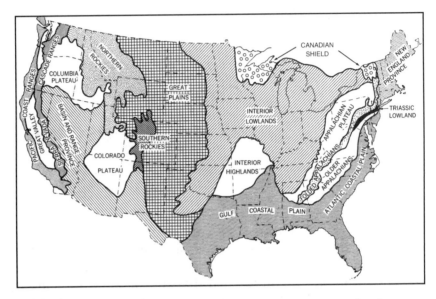

Figure 11-1. Landscape regions of the United States. Most land areas can be divided into a number of well-defined landscape regions. Each region is dominated by landforms resulting from geologic features and a relatively uniform climate.

Plains. Interior plains have elevations that are somewhat higher than those of coastal plains.

Plains are formed when sediments are deposited or when mountains are worn down over time. The resulting soil of plains regions is generally fertile. Both agricultural crops and animals, such as cattle and sheep, are raised on plains landscapes.

PLATEAUS

A **plateau** is a raised and rolling landscape confined on one or more sides by a steep boundary. Plateaus are higher in elevation than plains, but they are not as rugged as a mountain landscape. Plateaus rise at least 600 meters above sea level. The name *plateau* comes from the French for "flat," and in French *plateau* also means tabletop. Some locations, such as the region around the Grand Canyon in northern Arizona, are relatively flat. Most plateaus, however, are rolling or hilly landscapes. The Appalachian Plateau of the eastern United States is a rolling landscape.

Plateaus are found on every continent and in most oceans as well. Almost half of Earth's land surface is made up of plateaus. Large plateaus are located in Africa and Asia. In North America, plateaus exist along the mountains of the West and to the west of the Appalachians in the East. Most plateaus are located inland, but some plateaus do occur along coasts. When this happens, the plateau ends in a cliff either at the edge of a coastal plain or directly next to the ocean.

Unlike the broad streams and rivers that flow through plains, streams and rivers in a plateau often cut deep valleys below the general level of the land. In the Grand Canyon, the Colorado River has cut a deep valley through the Colorado Plateau.

Plateaus are formed in many different ways. For example, in Chapter 10 you learned that lava plateaus form during certain types of volcanic eruptions. Other types of plateaus, called intermontane plateaus, are formed within mountain systems. In the United States, the words *basin* and *range* are used to describe these plateaus. Such regions can be seen between the Sierra Nevadas and Rockies in the western United States.

Some plateau regions are dry areas that are often used for grazing animals, such as cattle, sheep, and goats. In the western United States, plateau regions contain valuable coal and mineral deposits.

MOUNTAINS

A **mountain** is a landform that rises sharply above the surrounding area. Mountains generally have narrow tops, steep sloping sides, irregular surfaces, and heights of more than 610 meters. In other words, a mountain is a landform with high elevation and high relief. Mount Everest, which rises 8846 meters above sea level, is the tallest mountain in the world. Mount Mc-Kinley, which rises 6194 meters above sea level, is the tallest mountain in the United States.

Streams and rivers in mountain areas move very quickly. The steeper the stream slopes, the faster the water flows. Streams and rivers often carve V-shaped valleys into mountains. You will learn more about this process in Chapter 14.

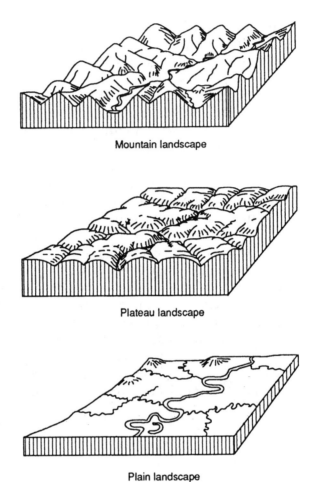

Mountain landscape

Plateau landscape

Plain landscape

Figure 11-2. Plains, plateaus, and mountains.

Most mountains are part of a group of mountains called a mountain range. A **mountain range** is a series of mountains that have the same general shape and structure. A **mountain system** is a group of mountain ranges in one area. The Appalachian Mountain system in the eastern United States contains the Blue Ridge, Great Smokey, and Cumberland mountain ranges. Most mountain ranges and systems are part of a larger group of mountains known as a **mountain belt**. There are two major mountain belts in the world. The Circum-Pacific belt runs along

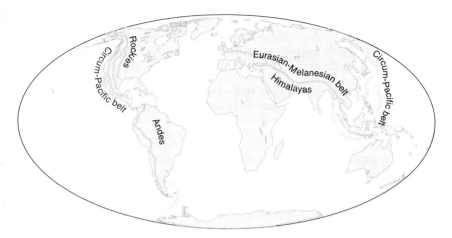

Figure 11-3 Two major mountain belts—the Circum-Pacific belt and the Eurasian-Melanesian belt—contain most of Earth's mountains.

the Pacific Ocean; the Eurasian-Melanesian belt runs across northern Africa, southern Europe, and Asia.

How Mountains Form

Mountains are built very slowly. For example, geologists estimate that the Rocky Mountains began to form about 65 million years ago. It took about 10 million years for these mountains to reach their maximum height. Even so, the Rocky Mountains are considered young mountains by geological standards. Other mountains have been forming for even longer periods. Over millions of years, fault movement can change a flat plain into a towering mountain range.

Mountains can form in a number of different ways. As you learned in Chapter 10, some mountains are formed from volcanic eruptions. Mount St. Helens in Washington is an example of such a mountain. The Cascades of the Pacific Northwest, in which Mount St. Helens is located, make up a string of isolated volcanoes.

Other mountains form from the folding and breaking of Earth's surface due to the movement of lithospheric plates. One type of mountain-building process occurs when oceanic crust collides with continental crust. When this happens, the heavier oceanic crust sinks, or subducts, under the continental crust. (Subduction was described in Chapter 8.) During this process, rock on the continent forms into mountains. The Andes Mountains near the west coast of South America were formed in this way. Beneath these mountains, the Nazca Plate is subducting under the continental South American Plate. (Refer to Figure 8-7.)

Another type of mountain-building process occurs when two continents collide. The Himalayas formed in this way as the continental crust of the Indian Plate collided with the Eurasian Plate. Before two continents can collide, oceanic crust subducts beneath one of the continents. Once the continents are in contact, subduction stops because continents are not dense enough to sink. Continued movement then causes the rocks on the continental margins to be crumpled into mountains.

When plates collide to form mountains, different types of mountains are produced. Some mountains are formed by faulting. Recall from Chapter 9 that a fault is a break or crack in Earth's surface along which movement has occurred. There are three basic types of faults. The San Andreas Fault you read about in Chapter 9 is a strike-slip fault. In this type of fault, the rocks on opposite sides of the fault plane move horizontally past each other.

A second type of fault is a reverse fault. This occurs when one side of the fault plane is driven up over the other side. This

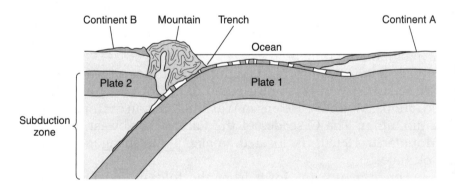

Figure 11-4. During subduction, the heavier oceanic plate sinks beneath the continental plate. This can result in the formation of mountains.

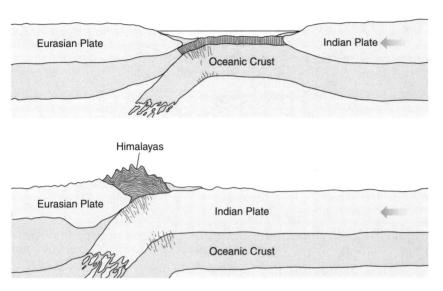

Figure 11-5. When two continents collide, rock is crumpled up into mountains. The Himalayas formed in this way.

type of faulting is important in mountains that are formed during plate collisions.

A third type of fault is a normal fault. This occurs when the rocks on one side of the fault plane drop down with respect to rocks on the other side. When two parallel normal faults exist, the block of rock between them can be pushed up or down as the two outside rocks move. If the inside block rises, a mountain is formed. If it falls, a valley is formed. The Teton Mountains of Wyoming are fault mountains where the mountain block has been pushed higher than the adjacent valleys. Since these faults occur where tension is pulling the crust apart, they are not a

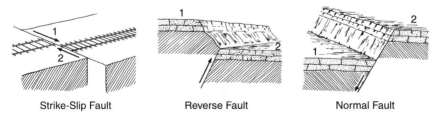

Strike-Slip Fault   Reverse Fault   Normal Fault

Figure 11-6. Three types of faults.

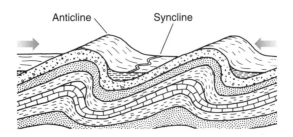

Figure 11-7. Plate movement can cause Earth's crust to fold. Some of the world's largest mountain ranges have formed in this way.

major force in mountain building. Mountains usually form where two plates are coming together.

Other mountains are formed by folding when rock layers are crumpled during plate collisions. The Alps in Europe formed when pieces of the crust folded during the collision of two plates. The Appalachian Highlands of North America are also fold mountains. Geologists use the terms *anticline* and *syncline* to describe upward and downward folds in rock. An anticline is a fold in rock that bends upward into an arch. A syncline is a fold in rock that bends downward in the middle to form a bowl.

QUESTIONS

Multiple Choice

1. A plains landscape is likely to have   *a*. low relief and low elevation   *b*. high relief and low elevation   *c*. low relief and high elevation   *d*. high relief and high elevation.

2. A mountain landscape is likely to have   *a*. low relief and low elevation   *b*. high relief and low elevation   *c*. low relief and high elevation   *d*. high relief and high elevation.

3. A raised and rolling landscape that has a steep boundary is most likely   *a*. a mountain   *b*. an interior plain   *c*. a coastal plain   *d*. a plateau.

4. A landscape in which rivers and streams cut deep channels is most likely   *a*. a plateau   *b*. a mountain   *c*. an interior plain   *d*. a coastal plain.

5. The largest grouping of mountains is a   *a.* system
   b. range   *c.* belt   *d.* valley.

Fill In

6. The _____ of a region describes its height above sea level.
7. A landscape that is high in some parts but low in others has high _____.
8. A landscape that gets its name because it resembles a table is a(an) _____.
9. A group of mountain ranges in an area is known as a mountain _____.
10. A(an) _____ fault, in which one side of the fault plane is driven up over the other side, is important in mountain building.

Free Response

11. Describe the general characteristics of plains, plateaus, and mountains.
12. How might you differentiate among the three different types of landscapes based on the rivers that run through them?
13. What are two ways mountains are formed when plates collide?
14. What are the three types of faults? How are they related to mountain building?
15. What happens when a mountain is formed by folding?

Portfolio

I. On a blank map of your state, show the major landscape regions.
II. Identify and photograph places in your community where the shape of the land has been changed by builders and developers. Make a poster showing the places before and after they were changed.
III. Assemble pictures from magazines, newspapers, or your own collection to illustrate different types of landscapes.

Chapter 12
Weathering and Soils

THE ROCKS THAT COVER much of Earth's surface may seem timeless and unchanging. Yet they are not. Perhaps you have seen photographs of the Sphinx in Egypt or Cleopatra's Needle in New York. The Sphinx was carved out of limestone. Cleopatra's Needle was carved from granite. Both structures have been worn away over time. What happened to them? Rocks are formed under conditions of high pressure and temperature deep within Earth. On the surface, however, rocks are exposed to an entirely different set of conditions. These conditions wear away the rocks over time. In this chapter, you will find out what factors affect rocks on Earth's surface and how rocks change over time. You will also find out how rocks form soil and why soil is important.

WEATHERING

On Earth's surface, rocks are exposed to water, wind, oxygen, and carbon dioxide. These and other factors cause rocks to break down. The process in which rock is broken down due to exposure to the atmosphere on Earth's surface is known as **weathering**. During weathering, rock is broken into smaller and smaller pieces.

Weathering involves many different processes. These processes are classified as either mechanical or chemical. **Mechanical weathering** occurs when rock is physically broken into smaller pieces. The composition of the smaller pieces is the

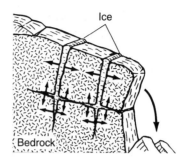

Figure 12-1. When water freezes, it
expands. If the water is in a crack in a
rock, it can push off a piece of the rock.

same as the original rock. Mechanical weathering occurs in
many different ways. One way is when rock is alternately exposed to temperatures below and above the freezing point of
water. Water is held in cracks and pores within rocks. When it
freezes, water expands, pushing apart rock. This process is
known as ice wedging or frost action.

Mechanical weathering also occurs when rock that contains
clay repeatedly becomes wet and then dries. When clay becomes
wet, it swells. Clay shrinks as it dries. Continual swelling and
shrinking causes rocks that contain clay to fall apart.

Living organisms also cause mechanical weathering. Lichens and mosses grow in tiny pores and cracks on rocks. In
time, rocks split as the lichens and mosses grow. In a similar
fashion, organisms such as ants, earthworms, and woodchucks
dig holes in soil. The holes allow air and water to reach protected rock, thereby weathering it.

Abrasion is another method of mechanical weathering.
Abrasion refers to the grinding away of rock by other rock particles carried in water, wind, or ice. For example, in fast-moving
rivers, bits of rock carried in the water collide with rocks in the
river's path. Over time, these rocks are weathered into small particles as well. In addition, the particles carried in the water are
further weathered. As particles are transported downstream,
they become smaller as a result of abrasion.

Chemical weathering occurs when the minerals within a
rock are chemically changed into different substances. Chemical weathering produces rock particles that have a different
composition from the rock from which they came. For example,
the weathering of feldspar in the presence of water creates clay.
In fact, a wide variety of minerals change into the family of

minerals that we know as clay. The agents of chemical weathering include water, oxygen, carbon dioxide, and living organisms.

Many minerals chemically react with water. When this happens, the rock breaks down into different substances. Other minerals chemically react with oxygen. You may know that an object made of iron will rust if it is exposed to air because oxygen in the air combines with iron to produce rust. In a similar reaction, rock that contains iron also rusts. This reaction makes the rock soft and crumbly and gives it a red or brown color.

Carbon dioxide can cause chemical weathering when it dissolves in water. During this process, a weak acid called carbonic acid is produced. The acid attacks many common minerals. Carbonic acid is particularly effective on calcite, dissolving it completely. Beautiful underground caverns have been hollowed out by the dissolving action of carbonic acid on calcite (the mineral in limestone). Some cave systems, such as Carlsbad Caverns in New Mexico, are famous for their unusual limestone formations. Similarly, several hundred miles of connected underground passages have been surveyed at Mammoth Caves National Park in Kentucky.

Living organisms are also agents of chemical weathering. Plant roots, for example, produce weak acids that slowly dissolve rock around the roots. In addition, acids are formed during the decay of dead plants and animals. These acids are dissolved by

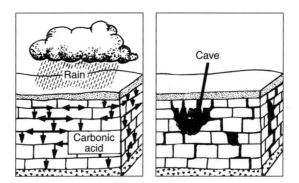

Figure 12-2. When carbonic acid seeps into limestone, beautiful caves are formed through chemical weathering.

rainwater and are carried through the ground to rocks, which are broken down by the acid.

RATE OF WEATHERING

Some types of rocks weather more quickly than others do. The minerals that make up the rock determine how fast it weathers. For example, shale is composed primarily of clay, a relatively soft substance that weathers quickly. Granite is composed of harder minerals, so it weathers more slowly. In places where different kinds of rock are found, the hills are composed of the harder types of rock and the valleys form in the softer rocks.

The rate of weathering is also related to the amount of rock exposed to the elements. When rock is broken, the surface area of the rock is increased. The greater the surface area, the faster the rock weathers away.

Climate is another important factor in determining the rate of weathering. Warm, moist climates support chemical weathering. Cold climates that have many cycles of freezing and thawing promote mechanical weathering in the form of ice wedging.

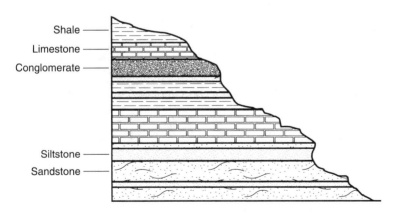

Figure 12-3. Some rocks weather more easily than others. Shale is relatively soft.

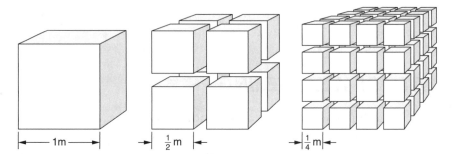

Figure 12-4. As a rock is weathered into pieces, its exposed surface area increases. This additional surface area accelerates the rate of weathering.

THE FORMATION OF SOIL

You may think of weathering as a destructive force, wearing down mountains and other landforms over time. But weathering also produces a valuable resource without which we cannot live—soil. **Soil** is a mixture of weathered rock particles and or-

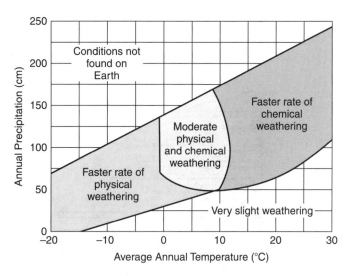

Figure 12-5. The type of weathering depends largely on the local climate. In general, chemical weathering dominates in warm moist climates, whereas cooler climates favor physical weathering.

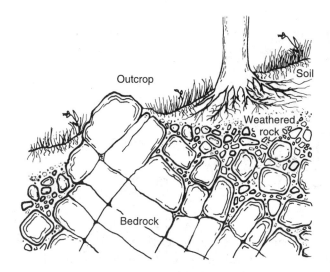

Figure 12-6. Rocks gradually weather over time to form soil.

ganic material. Plants with roots grow best in soil. Soil forms as rock is broken down by weathering and mixes with other materials on the surface. Soil is constantly being formed wherever bedrock is exposed. Bedrock is the solid layer of rock beneath the soil. Once exposed at the surface, bedrock gradually weathers into smaller and smaller particles that are the basic material of soil.

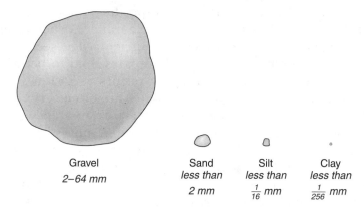

Gravel	Sand	Silt	Clay
2–64 mm	less than 2 mm	less than $\frac{1}{16}$ mm	less than $\frac{1}{256}$ mm

Figure 12-7. Comparison of soil particles by size.

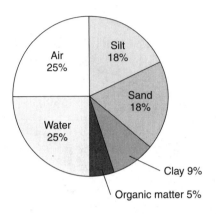

Figure 12-8. This pie chart shows the composition of loam.

The portion of soil from weathered rock is made up of silt, sand, and clay. The amounts of these rock particles determine the texture of the soil. Soil can be coarse and grainy or smooth and silky. Soil particles range in size from gravel, the largest, to clay, the smallest. Soil made up of about equal parts of clay, sand, and silt is called **loam.** Loam is best for growing most types of plants.

The decayed organic matter in soil is called **humus.** Humus forms as plant and animal remains decay. Humus is rich in nitrogen, sulfur, phosphorus; and potassium, which plants need to grow.

The nature of a soil is a result of three factors: bedrock, climate, and time. The raw material for most soils is bedrock. Bedrock is the layer of rock beneath the soil. Bedrock with a wide variety of minerals is likely to produce a fertile soil. Bedrock that consists primarily of a single mineral, such as calcite or quartz, is deficient in important nutrients. The second factor is the local climate. Soils in cold climates tend to be thin and rocky because chemical weathering occurs very slowly in cold climates. Soils in warm, moist climates are often very thick but lack important minerals because they are extensively weathered and leached. (Leaching means that water moves through the soil, carrying away minerals that dissolve in water.) Moderate or temperate climates generally produce the best soils. The final factor is the amount of time the soil has for development. A thick, fertile soil forms over thousands of years. Immature soils are not good for agriculture because they are rocky with little organic material.

SOIL HORIZONS

As mature soils form, they develop three distinct layers known as soil horizons. A **soil horizon** is a layer of soil that differs in color and texture from the layers above or below it. Soil horizons are generally called the A-, B-, and C-horizons.

Near the surface is topsoil. Topsoil is usually dark in color because it contains the most organic materials. Just below the topsoil is a zone of leached soil that is likely to be deficient in both organic content and minerals. These two layers together make up the A-horizon. The B-horizon begins with subsoil, which is a heavy reddish-brown soil made mostly of clay. Soil in this horizon is poor in organic material but enriched with minerals that have leached out of the A-horizon. The C-horizon is the deepest layer of soil. It is composed of broken bedrock with little chemical alteration or organic material. The rock fragments at the bottom of this horizon are on top of solid bedrock.

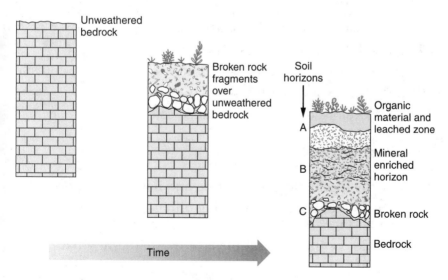

Figure 12-9. The development of soil horizons.

QUESTIONS

Multiple Choice

1. Mechanical weathering occurs when *a.* iron rusts
   b. carbon dioxide dissolves in water *c.* dead plants decay
   d. water freezes in cracks in a rock.

2. Chemical weathering occurs when *a.* a small plant
   wedges its roots into a rock *b.* an earthworm digs holes
   in soil *c.* plant roots produce weak acids that dissolve
   rock *d.* ice freezes into cracks in a rock.

3. Limestone caverns are formed underground when
   a. oxygen dissolved in water attacks calcite *b.* carbon
   dioxide dissolved in water attacks calcite *c.* iron is
   exposed to oxygen *d.* bedrock is exposed to air.

4. The smallest soil particle is *a.* sand *b.* silt *c.* clay
   d. gravel.

5. Which soil horizon is closest to bedrock? *a.* A *b.* B
   c. C *d.* D

Fill In

6. _____ is the process in which rock is broken down by
   water, wind, oxygen, and carbon dioxide.

7. The solid layer beneath soil is called _____.

8. _____ is soil made up of equal parts of clay, sand, and silt.

9. Decayed organic matter in soil is called _____.

10. Soil develops distinct layers called soil _____.

Free Response

11. What is mechanical weathering and how might it occur?

12. How does chemical weathering differ from mechanical
    weathering?

13. How does climate affect the type of weathering that
    occurs?

14. Explain how soil is related to weathering.

15. Describe the layers of mature soil.

Portfolio

I. Compare a fresh sample of rock, recently exposed by road construction, with another piece of the same rock that has been exposed for a long time.

II. Relate the weathering of cemetery headstones with their age. (Be sure to get permission to enter the cemetery.)

III. Observe changes caused by weathering on objects outside, such as buildings, cars, and roads. Classify these changes as physical or chemical weathering and explain your reasoning.

IV. Find a location where a soil profile is exposed by erosion or construction. Try to identify the soil horizons.

Chapter 13
Erosion and Deposition

LONG AGO, the region now known as the Grand Canyon consisted of the Colorado River flowing over a flat plateau. Over time, the river weathered the rock beneath it and carried it away. As rock was continually broken down and removed, the Colorado River gradually chiseled out the awesome canyon that exists today.

The Grand Canyon was the result of two processes. First the rock was weathered. Then it was carried away. Weathered rock will stay in place unless the forces of nature carry it away. In fact, the sediment found in most places shows signs of having been moved. **Erosion** is the transportation of sediment by gravity, wind, or running water. No matter how the sediment is carried, it eventually settles in a new location. The process by which sediment is deposited, usually because the agent of erosion is no longer capable of transporting it, is known as **deposition**. Most sediment goes through many cycles of being picked up by an agent of erosion, being transported to a new location, and being deposited again.

In this chapter, you will learn about erosion and deposition by gravity and wind. In the following several chapters, you will encounter erosion by running water in the form of runoff, groundwater, glaciers, and oceans. Throughout these chapters, you will discover how the processes of erosion and deposition work together to shape and reshape Earth's surface over time.

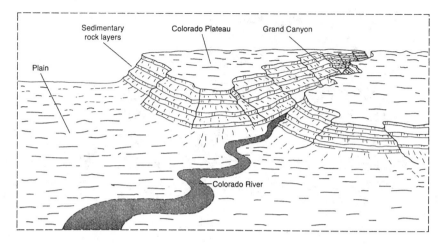

Figure 13-1. Many sites of natural beauty, such as the Grand Canyon in Arizona, formed as a result of erosion.

MASS MOVEMENT

Gravity is the force that pulls objects toward Earth's center. Gravity causes all objects near Earth's surface to fall to the ground. Weathered sediment, too, is pulled downward by the force of gravity. If the ground slants or slopes, the force of gravity will cause soil and rocks to fall to lower levels. Several processes through which sediment is pulled downhill due to gravity are called **mass movements**.

Examples of mass movements are landslides, slump, mudslides, and creep. During a **landslide**, dry rock and soil slide quickly down a steep slope. In a **slump**, a mass of rock and soil suddenly slips down a slope. Unlike a landslide, the material in slump moves down in one large mass. Slump often occurs when water soaks the base of a mass of soil that is rich in clay. It is common on steep slopes. In a **mudflow**, clay and silt that are saturated with water flow rapidly downhill. Mudflows often occur after heavy rains in normally dry areas. **Creep** is the slow downhill movement of rock and soil. It is so slow that it is imperceptible and occurs often on gentle slopes. You can tell that

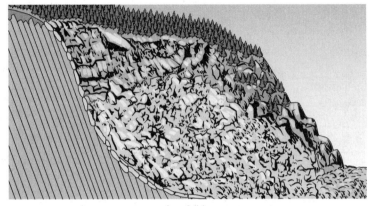

Landslide

Slump

Creep

Figure 13-2. Landslides, slump, and creep are three types of mass movement.

creep has occurred because it causes objects fixed in the soil to lean downhill.

When rocks fall from a steep cliff, they are deposited in a pile at the bottom of the cliff. If another agent of erosion does not carry them away, they remain in a disorganized pile. The particles in these deposits are usually very angular and vary from tiny silt particles to large boulders. The larger particles often dominate and may show fresh surfaces where the rocks recently broke apart. There is no separation by size or layers.

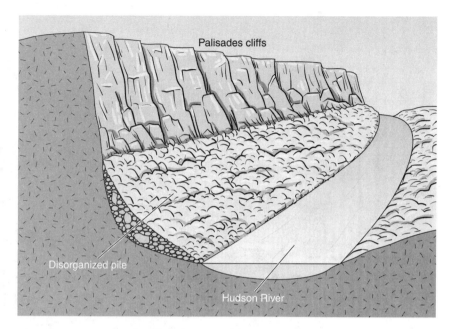

Figure 13-3. Rocks that fall from a cliff are not organized into layers.

WIND

Of all the agents of erosion, wind is the weakest. However, wind can be a powerful force in areas where sands, silts, and clays are loose and dry or where few plants hold the soil in place. For example, wind is very effective in deserts, where it can easily move the grains of dry, lightweight sand and silt.

Wind primarily causes erosion by deflation. **Deflation** is the process by which wind removes loose rock particles from Earth's surface. Wind sorts sediments by selective transport. This means that fine particles are picked up and carried by the wind. Larger pebbles and boulders are left behind. In places where the fine particles have been carried away by the wind, a stony cover known as desert pavement is left behind.

Some sediments, such as sand grains, cause further weathering through abrasion as they are carried by wind. Grains of sand, for example, are thrown against any rock in the path of the wind. Over time, this process can wear the rock down until it is

Figure 13-4. Desert pavement results when wind removes the sand, leaving behind flat, polished stones.

level with the ground. If part of the rock comes loose and is moved, a new face may be flattened. As a result, wind-weathered rocks tend to form flat faces. Coarse-grained rocks may also become pitted where the softer minerals are worn away more quickly than the rest of the rock.

The sediment picked up by the wind is eventually deposited when the wind slows or becomes trapped. This process of depo-

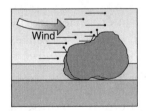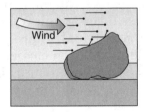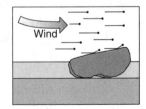

Figure 13-5. As sand is blown by wind, it grinds away the surface of rocks in its path.

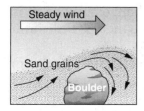
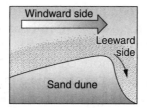
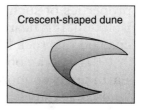

Figure 13-6. Windblown sediment can build up to form dunes.

sition can result in dramatic landscape features. For example, windblown sand sometimes forms sand dunes. Different shapes of sand dunes occur depending on the supply of sand, the wind strength, and the amount of vegetation. Sand dunes form on beaches and in some desert areas where windblown sediment has built up over time.

Sand dunes are usually made of the coarser sediments carried by wind. The finer sediments, called **loess**, are then deposited in layers far from their source. Loess, which helps to form fertile soil, consists of layers of fine, yellow-brown silt. Loess covers large areas of the central United States. For example, loess can be found in thick layers (tens of meters deep) that form steep bluffs along the Mississippi and Missouri rivers.

EARTH'S SYSTEMS

In Chapter 11, you learned about the different types of landscapes and how they formed over time. Just as various natural forces cause landscapes to build up, the forces of erosion cause landscapes to wear down. If building forces are dominant in a region, as in parts of the Himalayan Mountains of Asia, the land rises. If erosion is dominant, as in many agricultural areas of the United States, elevations generally decrease. If building and erosion are in equilibrium, land levels remain about the same.

It is important to recognize that many different forces are at work in any one region because Earth's systems are not closed systems. Isolated objects and systems are relatively easy to study and predict. An example is a rock deep in interstellar space. The properties of such a rock are unlikely to change because of its isolation. It is nearly a closed system because there is little influence on this rock from anything else.

Earth as an object is more complex. Objects and systems on Earth are even more complex. In this chapter, you learned that landscapes are constantly changing. They are continually reshaped by agents of erosion. In addition, the landscape is changed by human needs and technology. The human influence is probably the most unpredictable agent of change. Earth and its systems can be described as open systems. Open systems are influenced by outside objects and exchange both energy and matter with their surroundings.

Most systems that you will encounter in your study of earth science are open systems. Open systems are interesting because they are dynamic. But they are also difficult to study because of their complexity. This makes them difficult to predict.

QUESTIONS

Multiple Choice

1. Which choice is closest in meaning to *erosion*?   *a.* weathering   *b.* transportation   *c.* chemical change   *d.* physical change

2. The opposite of erosion is   *a.* motion   *b.* sediment   *c.* deposition   *d.* traveling.

3. Desert pavement is created by erosion due to   *a.* wind   *b.* water   *c.* gravity   *d.* ice.

4. How can we recognize rocks that have been eroded by wind?   *a.* They are often rounded with smooth surfaces.   *b.* They often have flattened and pitted surfaces.   *c.* They usually have smooth surfaces with grooves in them.   *d.* They usually form rectangular solids with smooth surfaces.

5. Which is the most common land feature in desert regions?   *a.* great areas of sand dunes   *b.* dense forests of tropical plants   *c.* flat land covered by sand   *d.* stony surfaces with a rocky cover

Fill In

6. Sediment is transported by the process of _____.

7. Sand dunes are created by _____, during which wind drops weathered particles.

8. The agent of erosion for landslides and mudslides is _____.

9. Trees on a hillside may lean downhill due to a mass movement known as _____.

10. Steep bluffs made up of _____ are composed of fine sediments deposited by wind.

Free Response

11. How is erosion related to weathering?

12. What is deposition?

13. How can gravity act as an agent of erosion?

14. Why are sediments deposited by wind more organized than those deposited by mass movement?

15. Why don't mountains continue to rise forever?

Portfolio

I. Make a pile of dry particles of mixed sizes and observe which are moved by the force of a fan or by a strong wind outdoors.

II. Measure the speed of the wind at the same time each day for a week. Make a histogram of wind speed and frequency.

III. Go outside and look for evidence of mass movement. Describe what type of mass movement you see and explain its possible causes.

Chapter 14
Earth's Rivers

DO YOU EVER watch heavy rainstorms and wonder what happens to all the water? When rain falls to the ground, some water sinks into the ground, some evaporates or is taken up by plants, and the rest moves over the land as runoff. **Runoff** is water that moves over Earth's surface. Runoff also comes from melting snow or ice. The moving water in runoff is the major agent of erosion on Earth. In this chapter, you will learn how runoff moves over Earth's surface and how runoff causes erosion and deposition.

PARTS OF A RIVER

When rain falls or snow melts, the resulting water flows downhill due to gravity. At first, the water forms tiny grooves in the soil called **rills**. As the rills flow into one another, they grow into gullies. A **gully** is a large groove in the soil that carries runoff after a rainstorm. Gullies join together to form a larger channel called a **stream**. As streams flow together, they form larger and larger bodies of flowing water. A large stream is often called a **river**. The beginning of a river, which is usually high in the mountains, is called the **headwaters**. The point at which a river flows into another body of water is called the **mouth** of the river.

The smaller streams and rivers that feed into a main river are called **tributaries**. A river and all of its tributaries make up a **river system**. The region of land drained by a particular river system is called the **drainage basin** or **watershed**. The high land between watersheds is called a **divide**. The longest divide

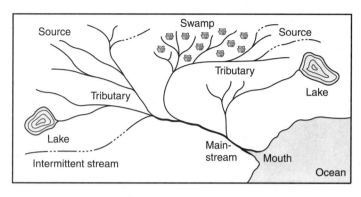

Figure 14-1. Parts of a river.

in North America is the Continental Divide along the Rocky Mountains. Rain that falls west of the Continental Divide eventually flows into the Pacific Ocean (or into the Great Basin where it evaporates). Rain that falls east of the Continental Divide eventually flows toward the Mississippi River or the Gulf of Mexico.

In general, the larger a watershed, the larger the river. There are several ways to determine the size of a river, but the most important measure is usually its discharge. **Discharge** is a measure of the volume of water flowing past a given location per unit time. It can be measured in cubic meters per second, or any other convenient combination of volume and time.

EROSION BY RIVERS

The channel over which a river flows is called the **riverbed**. As water flows rapidly over the rock of the riverbed, the water picks up rock fragments and carries them away. The particles in the water then cause further weathering through abrasion. Sand, pebbles, and other rock particles carried in the water act as cutting tools. These particles chip away at the riverbed as well as any other rocks in the river's path. At the same time, the particles are themselves worn down. Particles carried in rivers are usually much finer near the mouth of a river than near the headwaters because they are weathered along the entire length of the river.

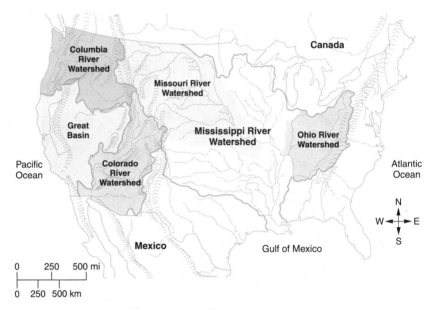

Figure 14-2. Major watersheds of the United States.

In addition to the physical processes just described, moving water causes chemical weathering by reacting with the bedrock to dissolve soluble minerals. During this process, holes are formed in the riverbed, and existing cracks and holes are widened. The products of these reactions are then carried away by the water. These products, which cannot be filtered out of the water, are carried in solution. (When a substance dissolves in water it is said to be in solution. Seawater, for example, is a solution of salts in water.) Compounds of calcium and magnesium are most commonly carried in solution.

Not all particles are carried in solution. The way that running water carries sediment is a function of the density and size of the particles being eroded. Small particles that are too tiny to be visible but are large enough to be filtered out of the water are carried in suspension. You may have observed a suspension if you have ever stirred the water in a puddle on a field or lawn. Although the water may have been clear, it became cloudy as the soil particles were lifted into the water. The same is true for particles of clay, silt, and fine sand in a river. The

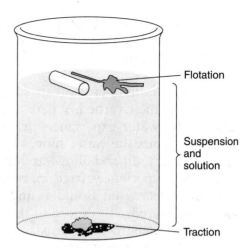

Flotation

Suspension
and
solution

Traction

**Figure 14-3. Water transports
(erodes) materials in four
ways: solution, flotation,
suspension, and traction.**

turbulence of the water stirs up these particles and keeps them
from sinking.

Large particles and those that are most dense are bounced
along the bottom of the riverbed in a process known as traction.
These particles move at a rate that is slower than the water in
the river. Particles that are less dense than water are carried on
the surface of the water by flotation.

ERODING POWER

The ability of a river to erode materials is directly related to
the slope of the river, its discharge, and the shape of the riverbed. The faster the water in a river moves, the more power it has
to erode the land beneath it and the more sediment it can carry.
The speed of a river increases with slope. Slope is the change in
the river's elevation over a distance as it travels. So the steeper
the slope, the greater the erosion.

The greater the discharge, the greater the eroding power. In
nature, discharge is especially important because it changes. A
sudden storm can increase the amount of water flowing in a river
or stream by many orders of magnitude. A small, intermittent
stream can change to a raging torrent in flood. Furthermore, the
discharge at the headwaters of a river is vastly different from the

discharge at the mouth. At the mouth of a stream, all of the water that was running through the river's tributaries is now flowing in the river. Therefore, the mouth has a greater discharge than the headwaters.

The amount of erosion also depends on the shape and composition of the riverbed. Where a river is deep, a small percentage of the water experiences friction due to contact with the riverbed. Thus the water moves faster and erodes the riverbed more quickly. In a shallow streambed, a greater percentage of the water experiences friction, reducing the river's speed. Thus erosion is decreased. Boulders and other obstacles in the streambed also decrease the speed of the river. In addition, a riverbed composed of hard rock resists erosion. Sometimes **waterfalls** occur where a river meets an area of rock that is very hard and erodes slowly. The river flows over this rock and then flows over softer rock downstream. The softer rocks wears away faster than the harder rock. Eventually, a waterfall develops where the softer rock was removed.

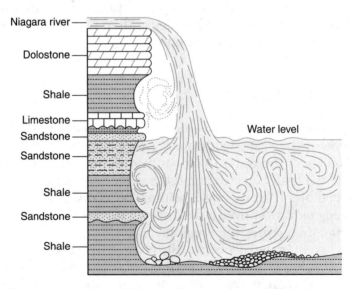

Figure 14-4. Waterfalls, such as Niagara Falls, erode their way upstream mostly by abrasion. Shale appears to be the weakest type of rock here, because it is eroding more quickly than the other rock types.

DEPOSITION BY RIVERS

The sediment in a river is carried until the characteristics of the river change. A river may deposit sediment as a result of a decrease in either speed or discharge. A river's speed decreases when its slope decreases, its bed widens, or it meets an obstruction. A river's discharge may decrease if it passes through an arid region where it loses water by evaporation into the air and seepage into the ground. Discharge my also decrease when people divert water for irrigation or water supplies.

Deposition from moving water creates several interesting landforms and changes the surface of the river valley. Where a stream flows out of a narrow mountain valley, the stream becomes wider and shallower. The decreased slope and increased width of the riverbed cause the water to slow down. As a result, the stream drops a large part of its sediment in a sloping, fan-shaped deposit called an **alluvial fan**.

During heavy rains, a river may overflow its banks and cover part of the valley floor. This part of the valley floor is called the **floodplain**. After a flood, both discharge and speed decrease as flooding waters subside. This causes the river to deposit sediment as new soil on the valley floor. The deposition of new soil over a floodplain is what makes a river valley fertile. After repeated floods, the floodplain becomes wider.

Floods are responsible for another type of formation known as natural levees. A natural **levee** is an elevated ridge made up of thick deposits built up alongside the river's banks. Behind the levee, the floodplain slopes away from the river. This allows swamps, called back swamps, to form in the lowest areas.

Obstacles in a river's path may cause the river to bend to one side or the other. When a river swings around a bend, the fastest-moving water is on the outside of the bend. Since erosion increases with speed, erosion is most rapid on the outside of the bend. The water moves more slowly on the inside of the bend. Sediment is often deposited there. Over time, the river eventually forms a series of broad curves called **meanders**. A meander is a looplike bend in the course of a river. Meanders become more and more curved as erosion occurs on the outside and deposition on the inside of the bend. If the river connects the ends

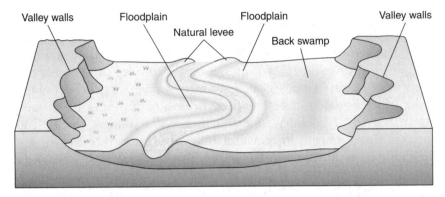

Figure 14-5. When a river overflows its banks, the sediment dropped from the water coats the floodplain and creates natural levees and back swamps.

of the meander, a crescent-shaped body of water called an **oxbow lake** is formed.

The greatest deposition occurs at the mouth of a river where the river empties into a larger body of water such as a lake, a gulf, or an ocean. Here the water slows dramatically, depositing most of its remaining sediment. Sediment that builds up at a river's mouth forms an area called a **delta**. The Mississippi, the Nile, and other great rivers have level, fan-shaped deltas.

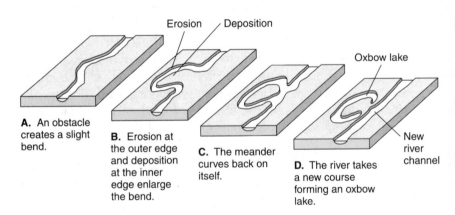

Figure 14-6. The formation of meanders and oxbow lakes.

The changes in a river are often described relative to the maturity of the river. A river is young at the headwaters where the slope is steepest and the water moves most rapidly. There the water erodes the land into a narrow, V-shaped valley. A mature river has a gentler slope and smooth riverbed. This happens because the river erodes the sides of the riverbed more than the bottom, creating a floodplain. An old river has several meanders and oxbow lakes as well as natural levees.

SORTING

The rate at which particles settle out of water depends on the size, shape, and density of the particles. Other factors being equal, the largest particles, the most dense particles, and the roundest particles are deposited first. If the particles are small enough, they can remain suspended indefinitely. Much like the tiny ice crystals or droplets of water that form in a cloud in the sky do not fall on a sunny day, neither does sediment held in solution or suspension. These smallest particles may stay in the water until evaporation occurs.

Sorting is the separation of particles on the basis of their properties. Among sediments, differences in the sizes of the particles are the most common indication of sorting. As a river slows, the largest particles settle first, and finer sediments are carried farther. This is known as **horizontal sorting**. Another type of sorting, **vertical sorting,** occurs when a mix of particles suddenly falls into deep water. This could occur from an underwater landslide or a cliff giving way into a deep lake. The largest, densest, and roundest particles settle first, followed by smaller and smaller particles. After the event that caused the particles to fall suddenly, a layer of sediment is left in which the largest particles are at the bottom with a gradual decrease in the size toward the top of the layer. This kind of gradual change within a single layer of sediment is called **graded bedding**. Each layer that grades from large particles at the bottom to smaller particles toward the top represents a single event.

CONCLUSIONS AND MISCONCEPTIONS

Making logical conclusions is an important part of the work of a scientist. If, for example, we see water just beginning to flow in a desert stream, we might conclude that there has been rain upstream. If we observe that some particles settle faster in water, we might conclude that those particles are probably larger, denser, and/or smoother than the particles that settle slowly.

Sometimes, however, people reach conclusions that are incorrect. For example, people once concluded that Earth must be flat and motionless because that conclusion supported their observations. However, Earth is so large we need to travel into outer space to easily observe its curvature.

Another misconception is the idea that the landscape is not changing. Although we can see that rivers sometimes flood and change their courses, it is rare to observe any major changes in the shape of the land. Most processes that change the shape of the land occur very slowly, and even a lifetime is not enough time to observe major natural changes in most landscapes. Yet we know that landscapes do change. Landscapes pass through stages of maturity, just as do the rivers they contain. The maturity of a landscape can be determined from the portion of the land that has been eroded down to base level. The maturation process can be thought of as mountains being worn down to featureless lowlands. Base level is the level at which the lowest streams flow from the land. For most locations near the oceans, base level is sea level. As you learned in this chapter, within the process of wearing down the land, the slope of the river decreases, and the bottom of the valley becomes wider. Rivers can develop meanders that continue to wear down the slopes at the edges of the valleys. The more mature a river, the more meanders it has and the wider its floodplain.

An incorrect conclusion is not a failure. It is an attempt to explain observations. However, the goal of a scientist is to continually ask questions and seek answers to those questions through research, experimentation, and analysis. Only in this way can we continue to develop and refine our understanding of the world.

QUESTIONS

Multiple Choice

1. The highland between watersheds is called a   *a.* rill
   b. gully   *c.* divide   *d.* drainage basin.
2. Deposition by running water is most likely to occur as
   the   *a.* water slows down   *b.* water speeds up   *c.* water
   becomes warmer   *d.* riverbed becomes steeper.
3. Which form of transportation moves sediment more
   slowly than the speed at which the water in river is
   moving?   *a.* solution   *b.* suspension   *c.* traction
   d. flotation
4. The least dense particles are carried in a stream by
   a. solution   *b.* suspension   *c.* traction   *d.* flotation.
5. Which factor does *not* influence the rate at which sediments settle out of water?   *a.* size   *b.* shape   *c.* hardness
   d. density

Fill In

6. The _____ is where a river begins.
7. The _____ of a river is the volume of water flowing past a
   given point per unit time.
8. Sediment deposited at the mouth of a river can form a
   fan-shaped _____.
9. _____ can be cut off by a river to form oxbow lakes.
10. Graded bedding forms from _____ sorting.

Free Response

11. What are the basic parts of a river system?
12. How do the size and density of a particle determine how it
    will be carried by a river?
13. Describe three factors that determine a river's ability to
    erode the riverbed.
14. How does a waterfall develop?

15. Explain how deposition by rivers changes the floodplain and forms natural levees.

16. How do erosion and deposition lead to meanders and oxbow lakes?

Portfolio

I. Find a place where stream erosion is active. Try to observe and record erosion in progress.

II. Collect samples of stream water both in flood and in low flow. Allow the water to evaporate. When does the stream carry the most sediment?

III. Observe the particles in a nearby stream to determine how their size is related to slope of the stream and the speed of the water.

IV. Make graded bedding by emptying a small beaker of sand of mixed sizes into a tall, transparent plastic column of water.

Chapter 15

Groundwater on Earth

WHERE DO YOU THINK most of Earth's usable freshwater is located? You may be surprised to learn that about 98 percent of the world's usable freshwater is in the ground. When rain falls and snow melts, some of the water evaporates, some becomes runoff, and some soaks into the ground. The water that soaks into the ground fills spaces between soil particles and cracks in rock. This underground water is called **groundwater**. In this chapter, you will learn how groundwater flows through soil and rock, why groundwater is important, and how groundwater causes erosion and deposition.

WATER IN THE GROUND

Water seeps into openings in the ground to become groundwater. Under certain conditions, however, water is unable to seep into the ground. For example, there may be no open spaces for the water to seep into, the surface may be so steep that the water runs off before it can infiltrate, the ground may be frozen, or the soil may already be saturated.

If water can seep into the ground in a particular region, the amount of groundwater stored depends on the spaces between soil particles. These spaces between soil particles are called **pores**. The **porosity** of a material is the percentage of its volume that is pore space. A sandy soil with grains of a uniform size can have a porosity of 33 percent or more. That means that the space available for groundwater storage is about one-third of the total volume of soil.

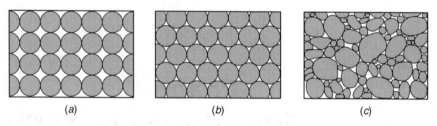

Figure 15-1. Porosity depends on the packing, shape, and sizes of particles. Different packing makes (*a*) more porous than (*b*). The porosity of (*c*) is reduced because it contains particles of mixed sizes.

The porosity of a soil depends on the shape and sizes of its particles. Rounded particles have more space between them than flat or angled particles. The pore space is also greater if the particles are all about the same size. If a material contains particles of different sizes, smaller particles can fill spaces between larger particles.

If the pores are connected, water can easily flow from one pore to another. If the pores are not connected, water cannot get into them. A material that allows water to pass through it easily is said to be **permeable**. **Permeability** is the rate at which water passes through the pore spaces of a material. Sand and gravel

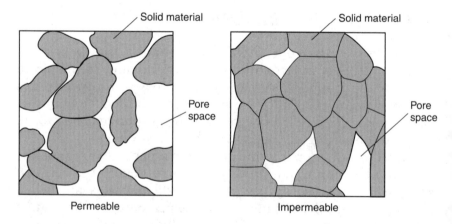

Figure 15-2. Water can move through the spaces in permeable materials. Impermeable materials do not have connected spaces through which water can move.

are examples of permeable materials. Materials are said to be **impermeable** if water cannot pass through them. Clay and granite are examples of impermeable materials.

When water soaks into the ground, it continues to move downward until it reaches an impermeable layer. Then, much like reaching the bottom of a glass, the water begins to fill the pore spaces above the impermeable layer. The water level in the ground rises as more pore spaces are filled. An area of the ground in which all of the pore spaces are filled with water is called the **zone of saturation**. The surface of the zone of saturation is called the **water table**.

Once in the soil, the movement of groundwater is driven by the same force that causes the movement of runoff—gravity. Groundwater always flows from a higher place to a lower place. Depending on the slope and the permeability of the soil, water moves at speeds from several meters per minute in sandy soils to several meters per year in dense soil.

An underground layer of permeable material that contains groundwater is called an **aquifer**. Aquifers can be small areas or they can be as large as several states. The Ogallala aquifer, for example, extends underground from South Dakota to Texas. Some aquifers form between two layers of impermeable rock. This arrangement is called an artesian formation. Rain that enters this aquifer is trapped. Water that moves into the aquifer is pushed along by the weight of the water behind it. The water continues to move as a result of gravity.

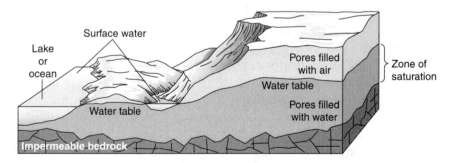

Figure 15-3. Most groundwater occupies the zone of saturation where pores are filled with water. It is bounded by the water table on top and impermeable bedrock below.

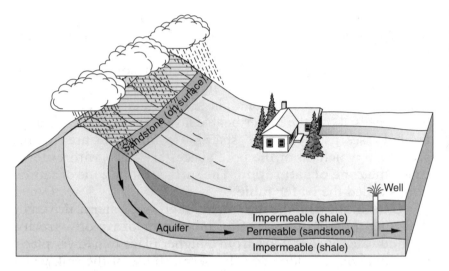

Figure 15-4. Rainwater that enters an aquifer of an artesian formation is trapped. The water may travel great distances underground before it rises to the surface through a well or spring.

GROUNDWATER RETURNS TO THE SURFACE

In many regions, groundwater is used for drinking water as well as for irrigation of agricultural crops. People can bring groundwater to the surface by drilling wells that reach below the water table. Some wells require pumps to bring the water to the surface. Artesian wells use the natural pressure created in artesian formations to obtain water.

In some places, the water table is already at the surface. In such areas the groundwater may feed streams, ponds, lakes, swamps, or oceans. Groundwater also comes to the surface as springs and geysers. **Springs** are places where groundwater flows out of cracks in rocks. A **geyser** is a spring from which heated water periodically shoots into the air. Geysers occur where heated rock is close enough to the surface to boil groundwater. Under the pressure of many meters of water above, water boils at an elevated temperature. As the superheated water makes its way to the surface, pressure is released, and an expanding volume of water suddenly boils. The result can be an intermittent spray of water and water vapor that surges tens of meters into the air before new water enters the heat source. The world's greatest region of geysers is in Wyoming's Yellowstone National Park.

GROUNDWATER CONSERVATION

Groundwater is an extremely important resource. Not only is it a major source of freshwater on Earth, in some places groundwater is the only water available. Even in dry climates where there is little or no surface water, it is often possible to find water deep within the ground. There are several advantages of groundwater. Unlike streams and rivers that can quickly go from flood conditions to drought conditions, the flow of groundwater is relatively constant. The small openings through which groundwater must pass even out the flow, reducing the need for surface reservoirs. The infiltration process also helps to filter and purify groundwater, which is usually safe for drinking.

Even though there is a great volume of water underground, supplies of groundwater are not endless. The local population may require more fresh water than they can obtain from the ground. This is especially likely where the precipitation is low and where water is needed for irrigation. If more groundwater is removed through wells than is returned by nature, the water table will drop. This may cause some wells to go dry. The Ogallala aquifer of the Great Plains is used for irrigation of crops from Colorado to Texas. Withdrawal has been much more rapid than the rate at which water can seep back into the ground (recharge). The result is that the water table is rapidly getting lower as the aquifer is depleted. Farmers using this source of water must either conserve water or plan for the time when it runs out. The problem is not limited to farmland. As cities grow larger, they too must find new sources of water. In addition to rising demands for water, land development replaces natural plant cover with buildings and pavement. This prevents water from reaching underground aquifers. Most large cities use rivers and reservoirs to supply their tremendous water demands.

In coastal regions where fresh groundwater floats on top of salt water, a removal of too much freshwater causes salt water to take its place. This causes wells to become unusable. On New York's Long Island, so much water has been drawn from the ground that salt water from the ocean is seeping into wells.

Another danger to groundwater is pollution. Since groundwater is a result of rain seeping through the ground, any pollutants in the ground will become part of the groundwater. This includes fertilizers, pesticides, and toxic chemicals. Even salt used to melt ice on highways seeps into the groundwater. Groundwater moves so slowly that contaminated areas remain polluted for hundreds of years. There is no easy or inexpensive way to purify polluted groundwater. At this time, the only way to protect groundwater is to prevent further pollution.

GROUNDWATER EROSION

Just as running water changes Earth's surface, groundwater also shapes the land through erosion and deposition. In this case, the result can be magnificent underground caves and caverns. If you have never toured a natural cavern system, a unique adventure awaits you. Caves are found in every region of the United States. Some caves occur in the jumble of rocks that fall from a cliff, some are made by the abrasive action of waterfalls or ocean waves, but the most extensive caves are carved by the chemical weathering of limestone. Recall from Chapter 12 that carbon dioxide in the air dissolves in rainwater to form a weak acid called carbonic acid. You may also recall from Chapter 5 that calcite, the mineral of which limestone is composed, can be identified because of its reaction with acids. Groundwater, which contains carbonic acid, therefore dissolves limestone. Over thousands of years, the acidic groundwater slowly dissolves the limestone and carries part of it away. This process gradually hollows out pockets in the rock. Over time, these pockets develop into large underground caves and caverns. If the water table drops, a dry cave large enough to walk through may be left behind. As mentioned in Chapter 12, the most exten-

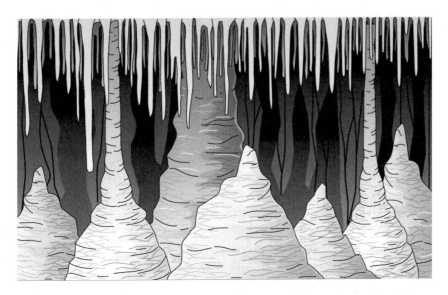

Figure 15-5. Stalactites and stalagmites form as a result of groundwater reacting with limestone.

sive system of underground caves known is Mammoth Caves in Kentucky, where many hundreds of kilometers of connected passages have been surveyed.

Water containing carbonic acid and calcium from limestone drips from a cave's roof. As the water evaporates, a deposit of calcite is left behind. A deposit that hangs from the roof is called a stalactite. A cone-shaped deposit on the cave floor is called a stalagmite.

If the roof of a cave collapses because of the erosion of the underlying limestone, the result is a depression called a **sinkhole**. Sinkholes often occur in regions where a layer of limestone is close to the surface. When this happens, water sinks easily into the ground, so streams are rare. Regions characterized by sinkholes and underground drainage are said to have karst topography. Karst topography is found in regions where the bedrock is made of minerals that dissolve easily. In the United States, there are regions of karst topography in Florida, Kentucky, and Indiana.

QUESTIONS

Multiple Choice

1. An example of a permeable material is   *a.* granite   *b.* bedrock   *c.* sand   *d.* clay.

2. If there is an increase in precipitation, the water table will   *a.* rise   *b.* drop   *c.* remain the same   *d.* move to a new location.

3. How deep must you dig a well to get a continuous supply of groundwater?   *a.* into Earth's mantle   *b.* just below Earth's crust   *c.* into solid bedrock   *d.* below the water table

4. Where is the world's greatest concentration of geysers and hot springs?   *a.* New York   *b.* Florida   *c.* Yosemite National Park   *d.* Yellowstone National Park

5. Which form of waste disposal is the greatest threat to the quality of groundwater?   *a.* burial   *b.* incineration (burning)   *c.* recycling   *d.* venting into the atmosphere

6. Which earth material can be identified by its bubbling reaction with acid?   *a.* quartz   *b.* limestone   *c.* mica   *d.* feldspar

7. Which change in state of water is necessary for the eruptions of geysers?   *a.* freezing   *b.* condensation   *c.* vaporization (boiling)   *d.* melting

Fill In

8. The groundwater in an area is related to the size and arrangement of the _____, the spaces between the soil particles.

9. A (an) _____ material allows water to flow through it easily.

10. A material is _____ if it does not allow water to flow through it.

11. The _____ is the surface of the zone of saturation.

12. The _____ is a region of permeable rock or soil containing water.

13. Regions of _____ topography can be found in Florida, where sinkholes and underground drainage are common.

Free Response

14. What happens to the water table when water is taken from an aquifer more quickly than it can be replenished?
15. Describe two conditions in which water will not seep into the ground.
16. What characteristics of a soil result in a high permeability?
17. What does it mean if a soil has a porosity of 25 percent?
18. What two conditions are needed for geysers to form?

Portfolio

 I. In a tall column of water, investigate how changes in density, size, and shape affect the time it takes for objects to settle to the bottom.
 II. Find out where your municipal water comes from and what your community does to protect its quality. Try to find out if your source of municipal water has changed in the past.
III. Prepare an exhibit of captioned pictures from a location where geysers and hot springs are active.
IV. Visit a nearby cavern system, either in person, through tourist information, or on the Internet. Your teacher may be able to help you find information.
 V. In some desert or island locations seawater is used for the public water supply. How is seawater made fit for drinking and other uses? Why is seawater used only in places where there is not enough surface water or groundwater?

Chapter 16
The Work of Glaciers

Some of the most spectacular lakes in the world can be found in the United States. These beautiful lakes may be surrounded by tall green trees, often with towering mountains in the distant background. What comes to mind when imagining such a lake? Perhaps you think about boating, having a picnic, painting a picture, or maybe a huge block of ice. Ice? Well, maybe ice isn't the first thought that comes to mind, but it should be. Many of America's lakes were formed through erosion and deposition by huge blocks of ice called glaciers. In this chapter, you will find out what glaciers are, how they are formed, and how they can change Earth's surface.

WHAT IS A GLACIER?

A **glacier** is a large mass of ice and snow. Glaciers form wherever the amount of snow that falls is greater than the amount of snow that melts. This usually occurs in high mountains where temperatures seldom rise above freezing. If this happens year after year, the snow builds to great depths. The lowest elevation to which the snow melts is called the snowline, and areas where snow lasts from year to year are known as snowfields.

If you squeeze a snowball, it will form a compact ball of ice. In a similar way, the weight of the snow in a snowfield changes the snow at the bottom into ice. When the snow and ice become thick enough, the mass begins to flow as a glacier. Although glaciers can advance many meters per day in temporary surges, glaciers more commonly move a few centimeters per day. At the end of a glacier, the ice melts as fast as it moves. This point, at which the glacier ends, is called the **ice front.**

TYPES OF GLACIERS

There are three basic types of glaciers. A **valley glacier** is a long, narrow stream of ice that creeps down a mountain valley. The sides of the valley keep these glaciers from spreading out in all directions. Valley glaciers are also known as alpine glaciers, named after the Alps of Europe where they were first studied. The largest alpine glaciers may be more than a hundred thousand meters long, several thousand meters wide, and hundreds of meters thick. The world's largest alpine glaciers are in southern Alaska.

If an alpine glacier reaches a low plain where the ice can spread out, it becomes a **piedmont glacier.** A piedmont glacier is a thick sheet of ice at the base of a mountain range. Large piedmont glaciers occur along the coast of Alaska.

Continental glaciers, or ice sheets, form in polar areas where the snow line is close to sea level and wide areas are above the snow line. Continental glaciers are much larger than valley glaciers. Today, continental glaciers cover about 10 percent of Earth's land. They are found in Greenland and Antarctica. The Greenland glacier covers almost 2 million square kilometers and is up to 3 kilometers thick. The Antarctic glacier covers about 13 million square kilometers and is more than 4 kilometers thick. The Antarctic ice contains about 75 percent of the world's freshwater. Continental glaciers are so large that they cover entire mountains.

EROSION BY GLACIERS

Although glaciers work slowly, they are major forces of erosion. Like rivers, glaciers remove loose rock from the land over which they move. Unlike rivers, however, the particles eroded by glaciers can range in size from fine powder to giant boulders. (Large boulders moved by glaciers are called **erratics**.) The reason is that rocks and rock fragments freeze into the

Figure 16-1. A valley glacier resembles a river of ice. When a valley glacier spills onto flat land, it forms a piedmont glacier.

bottom of the glacier. When the glacier moves, it carries the rocks with it.

The rocks frozen into a glacier scratch the bedrock as they are dragged over the land. Recall from Chapter 14 that this process is called abrasion. Fine particles act like sandpaper, smoothing and polishing the bedrock. Coarse particles leave long parallel scratches called **striations**. By studying striations, scientists can identify the general direction of ice movement.

Different types of glaciers change the shape of Earth's surface in particular ways. Erosion near the beginning of a glacier wears down the walls of mountain peaks. Over time, this process creates bowl-shaped basins called **cirques**. If several cirques are

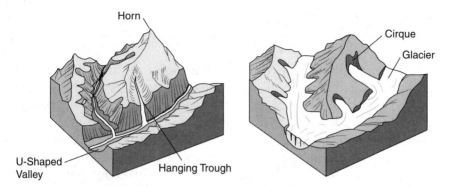

Figure 16-2. Mountain glaciers leave rounded valleys called glacial troughs. Hanging troughs are side valleys that end at a waterfall. Cirques are bowl-shaped depressions that form near the head of a glacier. The joining of several cirques results in a pointed formation known as a horn.

formed on the same peak, they wear the mountain into a point. These pointed peaks are called horns.

Alpine glaciers also create U-shaped valleys. Recall from Chapter 14 that fast-moving rivers carve out V-shaped valleys in mountains. Unlike rivers, which cover only a small portion of the valley floor, an alpine glacier covers the entire valley floor. It also comes in contact with the valley walls. So an alpine glacier wears away the floors and walls of the valley. The result is a glacial valley that has the shape of a U. This formation is called a **glacial trough**. Just as rivers are fed by tributaries, so too are glaciers. A U-shaped valley formed by a tributary glacier is called a hanging trough. Rivers that have formed in hanging troughs after the glaciers receded form spectacular waterfalls. One such waterfall is Yosemite falls in California.

Continental glaciers erode bedrock in much the same way as valley glaciers do. However, since continental glaciers cover most mountaintops, they wear away mountain peaks. Instead of shaping them into points, continental glaciers polish and round them.

DEPOSITION BY GLACIERS

When a glacier melts, it deposits the sediment it eroded from the land, creating various landforms. Deposits left behind by glaciers are called **drift**. Some deposits are left directly as debris drops out of the glacier. The sediments deposited directly by a glacier are called **till**.

Two types of till formations are moraines and drumlins. A **moraine** is a ridge of till deposited by a glacier as the glacier melts. There are several different types of moraines. A **ground moraine** is a thin layer of deposits that blanket an area when the rocks within the glacier fall to the ground. A ground moraine forms when the main body of a glacier melts. **Lateral moraines** form along the sides of a glacier. They form from rock fragments plucked from the valley walls. An **end moraine** forms at the end of a glacier, which is the ice front. An end moraine builds up at a point where the glacier melts as fast as it moves. When this happens, the glacier acts like a huge conveyor belt, moving rock particles toward the ice front. Even though the ice front does not move, rock continues to move forward. Since a glacier may stop in different locations for extended periods of time, more than one end moraine may form. A **terminal moraine** is the end moraine at the farthest point reached by a glacier. Long Island in New York was formed by two terminal moraines from past continental glaciers.

Unlike moraines, **drumlins** are made up of long, smooth hills of till often a hundred meters high and a kilometer long. Drumlins point in the direction in which the glacier was moving. Drumlins were probably formed when a glacier ran over an earlier moraine, sweeping it into a long strip.

Not all drift drops directly out of a glacier. Other drift settles out of water that melts from glaciers. Glacial deposits related to streams of melted water are called **outwash**. Glacial meltwater (water melted from a glacier) pours out at the ice front in streams filled with rock powder, sand, and gravel. The deposits from these streams look much like the alluvial fans of rivers. When the deposits overlap in front of large glaciers, they form broad, flat areas called **outwash plains**.

Other types of outwash include eskers and kames. These result from streams that form in and on a glacier. Streams form and run in tunnels that come out at the ice front. When the glacier melts, the deposits in these streams slump down at the sides, forming long, winding ridges called **eskers**. Eskers usu-

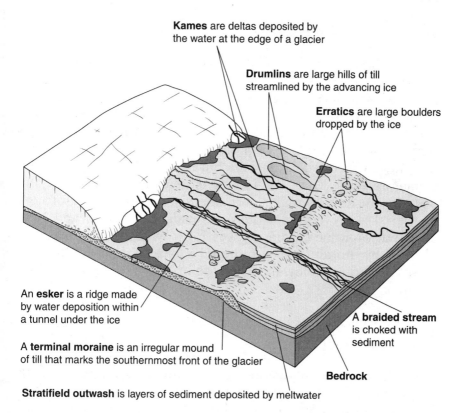

Kames are deltas deposited by the water at the edge of a glacier

Drumlins are large hills of till streamlined by the advancing ice

Erratics are large boulders dropped by the ice

An **esker** is a ridge made by water deposition within a tunnel under the ice

A **braided stream** is choked with sediment

A **terminal moraine** is an irregular mound of till that marks the southernmost front of the glacier

Bedrock

Stratifield outwash is layers of sediment deposited by meltwater

Figure 16-3. Continental glaciers leave a variety of characteristic features in the land.

ally run in the same direction as the glacier moved. When streams from the top of a glacier deposit their sediments along the edge of the ice or into lakes in the ice, they form **kames**. A kame takes the form a small, cone-shaped hill.

The drift deposited as till is different in its organization from drift deposited as outwash. Recall that the material carried by a glacier can be any size. Particles range from the finest clay particles to boulders as large as a house. Till is generally a jumble of unorganized and unsorted sediment because the collection of assorted debris is simply dropped as the ice melts. Outwash deposits, however, are sorted by size and separated into layers much like the sediment carried by rivers.

GLACIAL LAKES

The hollows and basins created by glacial erosion and deposition become lakes if they are permanently filled with water. Many of these lakes are common throughout the United States. Cirque lakes form when cirques left by valley glaciers become filled with water. Kettle lakes form in circular hollows in terminal moraines and outwash plains. These hollows, called **kettles**, are formed when deposits bury large blocks of ice as the glacier recedes. Then the ice melts, leaving the kettle. If the water in the kettle remains permanently, a kettle lake is formed. Kettle lakes are common in Wisconsin, New York, and New England. Moraine-dammed lakes form when river valleys are blocked by moraines. The moraine acts as a dam, forcing the river to flood its valley. Lake George in New York is a moraine-dammed lake.

THE ICE AGE

About a million years ago, the climate in much of North America and northern Europe was much colder than it is today. Huge ice sheets covered much of these areas. During the million-year period as the climate changed from cold to warm and back again, the ice sheets underwent four major advances and recessions. The last time the ice sheets receded was about 11,000 years ago. Many geologists think that we are now in a warm period that will be followed by a return of the ice sheets in another 20,000 years or so. Others think that this warm period will last for millions of years.

How do scientists know the Ice Age happened? Erosion and deposition by glaciers provide some evidence. For example, terminal moraines marked the southernmost point reached by some ice sheets. Terminal moraines stretch from New Jersey to Pennsylvania, Ohio, Indiana, and westward into the state of Washington.

In addition, geologists observed polished and scratched surfaces in northern regions that were unlike those in southern regions. Further, they found giant boulders that differed in composition from the local bedrock. At first, they concluded that a

Figure 16-4. The saber-toothed cat and the woolly mammoth are among the animals that lived in North America during the last Ice Age.

great flood had eroded the bedrock and moved boulders. A Swiss naturalist named Louis Agassiz, however, offered another explanation based on studies of existing glaciers in the Alps of Europe. He suggested that the observations were very similar to glacial erosion and deposition in the Alps. A number of geologists then concluded that great ice sheets must have covered Earth during an ice age in the past.

Most of our information about the ice ages has come from studies of land-based sediments and organic remains. In recent years oceanographers have discovered geochemical signatures in deep-sea sediments that provide precise information about ocean temperatures at different times in the past. From this method of analysis it has been shown that long-term cycles of warm and cool temperatures have taken place throughout much of geologic history. The main reason that the earlier glacial periods were not known until recently is that the sediments were destroyed or moved by more recent glacial events.

The reason for these widespread climatic changes remains a puzzle to most scientists. Many have tried to match the ice ages with a variety of cycles, including the precession (wobble) of Earth's spin axis, but none of the cycles seems to account for the onset of the ice ages.

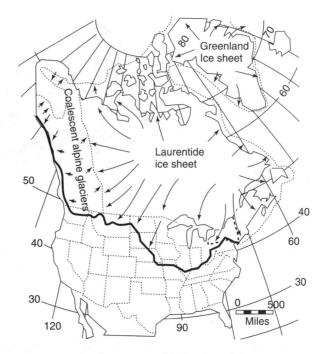

Figure 16-5. The geographic limits of the recent North American ice sheets.

THE CHANGING FACE OF GEOLOGY

Scientists have studied geology for hundreds of years. But the work of the geologist has changed in response to societal needs and emerging technologies. The earliest objectives of the geologist were to locate and estimate the available supplies of important resources, such as coal, copper, petroleum, and iron. These were the resources that spawned the industrial age and help to raise our standard of living. While pursuing these practical goals, geologists discovered that they could use a variety of features in the rocks to learn about ancient environments. From these beginnings, field geologists who study and map rock formations began to unravel the natural history of our planet.

Just as the goal of geology changed over time, emerging technologies changed the tools of geologists. By analyzing data

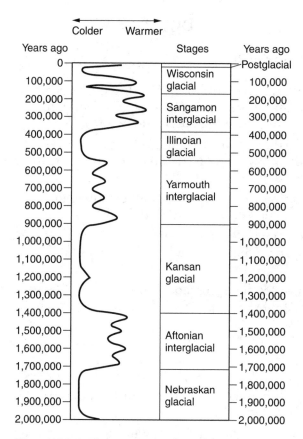

Figure 16-6. Fluctuations of world temperature over the past four glacial periods.

taken from airplanes and satellites, geologists have been able to measure quantities never before possible with a great deal of accuracy. Such measurements are particularly useful when it comes to motion that can be imperceptibly slow, such as the movement of glaciers. In addition, geologists are now able to use computers to create models that enable scientists to re-create conditions on Earth in the distant past and to predict condition in the distant future. Geology has come a long way from the simple equipment of the past—a rock hammer, a hand lens, and a bottle of acid.

QUESTIONS

Multiple Choice

1. A long, narrow stream of ice flowing in a mountain valley is   *a.* an alpine glacier   *b.* a continental glacier   *c.* a piedmont glacier   *d.* a cirque glacier.

2. The farthest point reached by a glacier is marked by *a.* a drumlin   *b.* a kettle lake   *c.* an esker   *d.* a terminal moraine.

3. A winding ridge deposited under a glacier is   *a.* a drumlin *b.* an esker   *c.* a moraine   *d.* a till.

4. Which type of moraine is formed when till is deposited along the edges of a valley?   *a.* terminal   *b.* end   *c.* lateral   *d.* ground

5. The last ice sheets receded about   *a.* 11,000 years ago *b.* 110,000 years ago   *c.* 1.1 million years ago   *d.* 1.1 billion years ago.

Fill In

6. The ice sheet covering much of Greenland is a(an) _____ glacier.

7. A(an) _____ glacier forms when a valley glacier spreads out at the bottom of a mountain.

8. Long, parallel scratches in bedrock called _____ are formed by the coarse particles in a glacier.

9. Long, smooth hills called _____ form from glacial till.

10. Large boulders called _____ do not match local bedrock if they have been transported long distances by ice.

Free Response

11. How do glaciers form?
12. Compare and contrast alpine and continental glaciers.
13. How does a river valley differ in shape from a glaciated valley?
14. Compare and contrast different types of moraines.

15. How has the climate of North America changed during the last two million years, and how do we know?

Portfolio

I. If you live in a region that you receives snow, measure the height of a snowbank and determine how quickly it melts away. Observe how the texture of the snow changes over time.

II. If you live in a glaciated region, make a list of local features produced by the glaciers. If you don't live in such a region, obtain photographs of such a region using books, magazines, and the Internet.

Chapter 17
The Restless Oceans

WHEN MOST PEOPLE think of Earth, they generally think of land. But more than 70 percent of the planet is covered by water. Earth is sometimes called the "blue planet" because all that water makes it look blue from space. Although the oceans are connected, geographers separate the vast region of water into four major oceans. The largest ocean is the Pacific Ocean, followed by the Atlantic Ocean. The third largest ocean is the Indian Ocean, which is located primarily in the Southern Hemisphere. The smallest ocean is the Arctic Ocean, which is near the North Pole. In this chapter, you will learn about the characteristics of the oceans, how ocean water moves, and how oceans are involved in erosion and deposition.

THE OCEAN FLOOR

Looking at the flat surface of the ocean, you may imagine the ocean floor as a flat surface as well. However, the ocean floor is not a vast, featureless plain. The lowest parts of the ocean floor are deeper than the tallest mountains on land. The ocean floor is commonly divided into two regions—the continental margin and the ocean basins.

Starting out from the shore, the continental margin begins with a gently sloping shallow area known as the **continental shelf**. This region is the part of the continent that is underwater. It extends to a depth of about 150 meters. After the continental shelf, the ocean floor plunges downward. This steep incline at the edge of the continental shelf is called the **continental slope**.

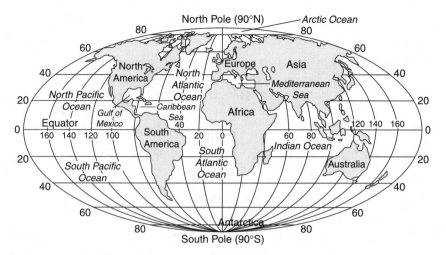

Figure 17-1. Most of Earth is covered by water.

Here the rock that makes up the continent stops and the rock of the ocean floor begins. The continental slope contains valleys that are as deep as the Grand Canyon. One such canyon is the Hudson Canyon, which extends from the mouth of the Hudson River in New York. The continental slope is followed by the **continental rise**. This area slopes gently to the ocean basin. In some locations, trenches exist instead of the continental rise.

The ocean basin is the floor of the deep sea. One interesting feature of the ocean basin is an abyssal plain. **Abyssal plains** are some the flattest areas on Earth. These plains are composed of sediment from the continents. The sediment can reach depths of more than six kilometers. Small rolling hills called **abyssal hills** are also found here. These hills often appear in groups near edges of continents or along ocean ridge systems. Cone-shaped mountains, called **seamounts**, are clustered near plate boundaries. Some seamounts, called **guyots**, have flat tops. They are thought to have risen above sea level and were then worn away by erosion.

The most obvious features of the ocean basins are the mid-ocean ridge and deep-sea trenches you read about in Chapter 8. Recall that the mid-ocean ridge is a continuous range of mountains that winds around Earth. Along the ridge new oceanic crust forms as two lithospheric plates move apart.

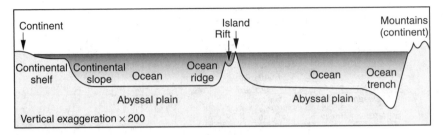

Figure 17-2. The ocean floor has several distinct regions.

THE COMPOSITION OF OCEAN WATER

Ocean water differs from freshwater because it contains dissolved salts. The most abundant substance dissolved in seawater is table salt (sodium chloride), but a number of other salts are also present. The salts in ocean water come from weathered rock. Most rock contains salts in very low concentrations. Over millions of years, most weathered rock changes to clay and silica that settle to the bottom of the oceans, and to salt. The salt dissolves in water and stays in solution. On average, 1 kilogram of ocean water contains about 35 grams of salt—35 parts per

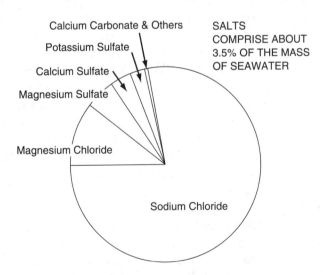

Figure 17-3. Ocean water contains several different dissolved salts.

thousand. The amount of salt dissolved in water is called the **salinity** of the water.

Overall, the salinity of ocean water remains constant over time. However, in certain areas, salinity increases. This is because water evaporates. When water evaporates, the materials dissolved in the water are left behind. So as ocean water evaporates, salt is left behind. The salinity of the remaining water is thereby increased. Salinity also increases near the poles where surface water freezes into ice. This, too, leaves the salt behind in the remaining water. In other areas, salinity decreases. For example, salinity decreases when freshwater is added to the ocean in the form of rain, snow, and melting ice or where large rivers empty freshwater into the ocean.

In addition to salts, ocean water contains dissolved gases. Marine organisms rely on these gases to survive, especially oxygen and carbon dioxide. Oxygen in seawater comes from the atmosphere and from algae that live in the ocean. Algae produce oxygen during photosynthesis. In the process, they use sunlight and carbon dioxide. Carbon dioxide is about 60 times as plentiful in the oceans as in the atmosphere. Animals, such as clams and corals, also use carbon dioxide in ocean water to build their hard shells and skeletons.

OCEAN TEMPERATURE

Ocean temperature decreases with depth. The reason for this is that almost all of the energy that heats ocean water comes from the sun. However, light and heat do not penetrate very deeply. Generally, ocean water is divided into three zones based on temperature: the surface zone of warm water with sunlight, a middle zone in which temperature changes rapidly, and a deep zone of very cold water and no sunlight.

The temperature of water also depends on latitude. Ocean water is warmer in the tropics than in other regions, and temperature drops as you travel away from the equator. In the tropics, where the noon sun is always high in the sky, sunlight is absorbed most efficiently. Near the poles, the sun is never high in the sky. When the sun is low, its rays must pass through more of the atmosphere, which absorbs some of the energy. That

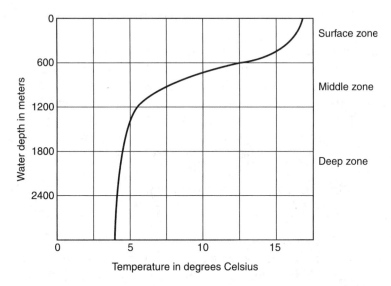

Figure 17-4. The temperature of ocean water changes with depth. This graph shows data taken near the equator.

leaves less energy available to reach the water. In addition, much of the ocean near the poles is covered with ice and snow. This causes energy to be reflected back into space.

The temperature of ocean water affects the amount of dissolved oxygen the water can hold. Cold water absorbs more atmospheric gases than warm water does. Thus cold water in the polar regions contains more dissolved oxygen than do warm, tropical waters. Phytoplankton, tiny plants that grow in the ocean, are at the base of the food chain in the oceans. These plankton thrive under conditions of abundant oxygen, which occur at the surface near the polar regions.

OCEAN CURRENTS

Ocean water moves in several different ways. One way is in the form of currents. A **current** is a large stream of moving water that flows through the oceans. Currents carry water great distances. Some currents move water at the surface. These currents, known as **surface currents**, are driven mainly by the wind and affect water to a depth of several hundred meters.

Surface currents move in a circular pattern that follows the major wind patterns. The currents move in circular patterns due to the rotation of Earth. If Earth did not spin, currents would flow in straight lines between the poles and the equator. But as Earth rotates, the paths of both the winds and currents are curved. This effect of Earth's rotation is known as the **Coriolis effect**. In the Northern Hemisphere, the Coriolis effect causes currents to curve to the right. In the Southern Hemisphere, the Coriolis effect causes the currents to curve to the left.

Other currents, known as **deep currents**, occur at the bottom of the ocean. Deep currents cause the cold waters at the bottom of the ocean to creep slowly across the ocean floor. These

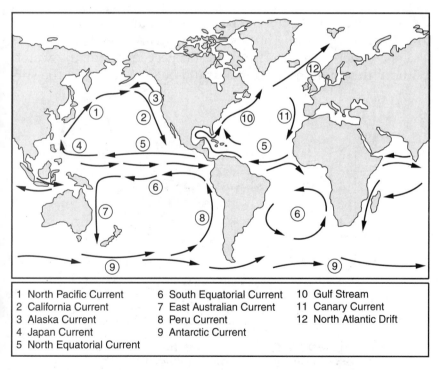

1 North Pacific Current	6 South Equatorial Current	10 Gulf Stream
2 California Current	7 East Australian Current	11 Canary Current
3 Alaska Current	8 Peru Current	12 North Atlantic Drift
4 Japan Current	9 Antarctic Current	
5 North Equatorial Current		

Figure 17-5. Circulation of the oceans is driven by uneven heating. The direction of flow is influenced by the positions of the continents, winds, and the Coriolis effect.

currents are caused by differences in density rather than surface winds. The density of water depends on temperature and salinity. Cold water is denser than warm water. And the higher the salinity, the greater the density.

When a warm surface current moves from the equator toward the poles, its water gradually cools. Thus it becomes denser. In addition, its salinity increases as ice forms near the poles. This too causes its density to increase. The greater density causes it to sink to the ocean floor. The cold, dense water then flows back toward the equator as a deep current.

Deep currents follow the hills and valleys of the ocean floor. Like surface currents, deep ocean currents curve due to the Coriolis effect. However, deep ocean currents flow more slowly than surface currents. It may take 500 years for water in a deep current to flow from the pole to the equator and back again.

In some places, cold water from ocean depths rises to the surface. This phenomenon, known as **upwelling**, occurs when winds blow away the warm surface water, and cold water rises to replace it. Upwelling brings up tiny ocean organisms, minerals, and other nutrients from the deepest layers of the water. Without this motion, nutrients would be very scarce in the sur-

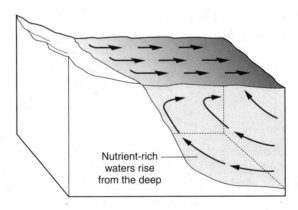

Nutrient-rich
waters rise
from the deep

Figure 17-6. Cold water from the deep ocean rises to the surface, bringing nutrients with it. This upward movement is known as upwelling.

face waters of the open ocean. Zones of upwelling are usually home to enormous schools of fish.

The flow of water in currents is a repeating cycle. In the tropics, water rises from great depth. As it does so, it brings nutrients and dissolved oxygen to the regions of the oceans that have the most abundant animal life. As the water rises, it is warmed by sunlight. It then flows away from the equator and toward the poles, where it becomes cooler, absorbs oxygen, and sinks. The cold water then flows back to the tropics, and the cycle repeats.

OCEAN TIDES

Another way that ocean water moves is in the form of daily tides. **Tides** involve the daily rise and fall of water along coastlines. In most places, the difference between high tide and low tide is about 1 meter. The world's highest tides occur in the Bay of Fundy in eastern Canada where they sometimes rise to 15 meters above the low tide level.

Tides are caused by the interaction of Earth, the moon, and the sun. Recall that gravity is the force that pulls objects toward each other. Every object in the universe exerts a force of gravity. The strength of that force between any two objects depends on the masses of the objects and the distance between them. The force increases with mass and decreases with distance.

The force of gravity exerted by the moon pulls on the water in Earth's oceans. The moon pulls most strongly on the water on the side closest to it. This pull creates a bulge of water, called a tidal bulge, on the side of Earth facing the moon. The water on the opposite side of Earth is pulled less strongly by the moon. This water forms a second bulge. High tides occur at the tidal bulges. Low tides occur in between them. As Earth rotates, different places experience high and low tides.

The sun also exerts a pulling force on Earth's waters. In Chapter 28 you will learn how Earth, its moon, and the sun move relative to one another. For now, it is enough to know that twice a month the sun and moon are lined up. When this happens, their combined gravitational pull on Earth's oceans produces the greatest range between high and low tides. These tides

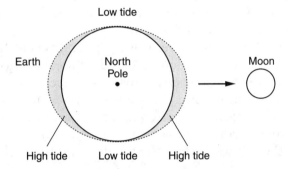

Figure 17-7. Tidal bulges form as a result of the moon's pull on Earth's water. Bulges form on the side of Earth closest to the moon and the side farthest away.

are called **spring tides**. In between spring tides, the sun and moon pull at right angles to each other. This arrangement produces a **neap tide**, a tide with the least difference between low and high tide. Spring tides and neap tides occur in a cycle that has a duration of about 29 days.

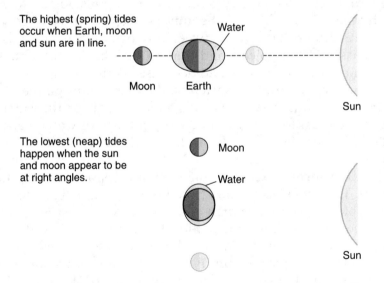

(Equivalent postions of the moon are lighty shaded.)

Figure 17-8. As the sun, Earth, and moon move relative to one another, spring tides and neap tides are created.

The movement of water between high and low tides is a source of energy that can be used to produce electricity. Tidal energy is a clean, renewable source of energy. However, there are very few places in the world where the height of the tides differs enough to make tidal power useful.

OCEAN WAVES

When you think of the ocean, you may imagine waves crashing onto the shore or raising surfers into the air. The third way that ocean water moves is in the form of waves. An ocean **wave** is an up-and-down movement of the water's surface. Ocean waves are most commonly produced by wind. Wind that blows for a long time or over a long distance creates waves. The size of the wave depends on how long the wind blows and the length of open water over which the wind blows. It also depends on the strength of the wind. Strong hurricane winds produce higher waves than normal winds.

In deep water, waves usually travel in long, low waves called **swells**. As the waves approach the shore, they enter shallow water. When this happens, the motion of the wave is changed. The bottom of the wave hits the ocean bottom and slows down. The top of the wave continues to move forward. When there is no longer enough water to support the wave, it falls over and crashes onto the shore.

You may think that waves move water, but, except for breaking waves, they do not. If they did, water would soon cover

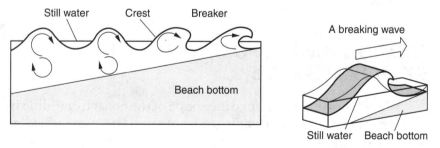

Figure 17-9. The sloping ocean bottom causes the bottom of an ocean wave to slow before the top. Eventually, the top crashes onto the shore.

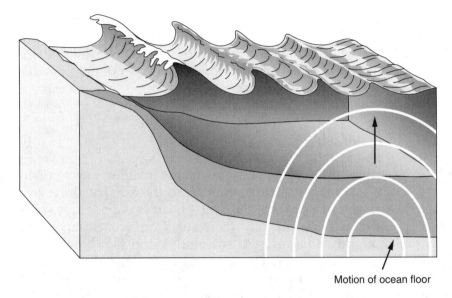

Motion of ocean floor

Figure 17-10. Movement of the ocean floor can create a tsunami. As the wave reaches the shore, its height can increase dramatically.

the land. Instead, waves transmit energy. (Recall from Chapter 9 that seismic waves transmit energy in a similar way.) The energy of ocean waves is transferred to the water by wind. Ocean waves then transmit the energy toward the shore.

Not all waves are caused by wind, however. **Tsunamis** are huge waves caused by earthquakes on the ocean floor. The movement of the ocean floor sends a pulse of energy through the water as a wave. Tsunamis can reach the shore as a wall of water taller than a five-story building.

EROSION AND DEPOSITION BY OCEAN WAVES

Waves continually change the shape of the coast by eroding it in some places and building it up in others. The most obvious weathering and erosion occur when waves break. When ocean waves reach the shore, they break fragments off of rock formations. The fragments are then carried away in the water. Breaking waves also force water into cracks in rocky cliffs. The pressure of the water can widen cracks and eventually split the rock. Rock fragments and sand carried in the water of the breaking waves further weather rocky shorelines. One more way that waves cause weathering and erosion is by dissolving minerals in rocks and carrying the products away.

The waves deposit sand, bits of rock, and pieces of shells along the shoreline to form beaches. The composition of beaches varies greatly. Some beaches are made of fine sand, whereas others consist of coarse sand. Some beaches are made up of rock, and still others are made of silt or mud.

Sand carried by waves and currents may deposit offshore as **sandbars**. If the sandbar grows well above sea level, it may form a **barrier island** parallel to the shore. Barrier islands and the shallow lagoons that form behind them protect the mainland from the full force of waves and storms. Beaches, sandbars, and barrier islands are constantly being changed as sand is removed and deposited by waves. Changes in the beach depend on the balance between erosion and deposition. If erosion dominates, as often happens in windy conditions and storms, sand is lost as the shoreline moves inland. If deposition dominates, the beach grows wider.

In places where beaches are wearing away or people have constructed buildings in vulnerable locations, various attempts to reduce erosion are in place. In some places **breakwaters**, piles of rock or concrete jetties, have been built into the ocean to protect the beach from wave erosion. These barriers often result in sand deposition on the windward (upwind) side, but sand is usually lost on the leeward (downwind) side. This occurs because most beaches are streams of sand flowing with the wind and waves along the shore. Extra sand that accumulates in one place is generally balanced by a loss of sand in another place.

Barrier islands and sand dunes are natural formations that reduce erosion. Barrier islands are helpful because waves break against the barrier beach instead of the land inside. Sand dunes are helpful because the roots of plants that grow in the dunes hold the sand in place. The plants thus reduce erosion by both wind and water.

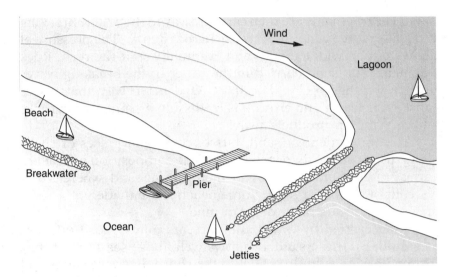

Figure 17-11. Structures, such as breakwaters, jetties, and piers, change the beach through erosion and deposition.

CONSERVATION OF OCEAN RESOURCES

In addition to a collection of minerals, the oceans contain a rich variety of plant and animal communities, especially in tropical reefs. Densely populated coastal nations such as Japan obtain a large percentage of their food supply from the oceans in the form of fish and bottom-dwelling animals. Extracts of ocean plants are also used as additives in foods, such as ice cream.

Despite their huge expanse, the oceans are not endless and they are not immune to abuse. People need to be aware of the oceans' resources so that they can conserve them. In the past, some valuable resources have been abused. For example, the most valuable species of whales have been hunted almost to extinction. Many varieties of fish are difficult to find because overfishing has dramatically reduced their numbers. Society must learn how to harvest the biological resources of the sea in such a way that the natural balance can be restored.

In addition to slowing the rate at which people *take* from the sea, they must slow the rate at which they *add* to the sea in the form of pollution. When a stream or a lake is polluted, the pollutants will eventually be washed away. But oceans are the final destination for most water pollutants. Evaporation leaves most pollutants behind in the seas, thereby increasing their concentrations. Mercury, a dense liquid metal often used in batteries and thermometers, is a pollutant that has been detected in the oceans. It is a dangerous toxin even in low concentrations. Mercury is one of a growing number of artificial toxic substances that have been measured in seawater. For years, people have dumped toxic wastes and garbage into the oceans thinking that they would not affect the huge amount of water—but they do.

QUESTIONS

Multiple Choice

1. Roughly what portion of Earth is covered by ocean?
   a. $\frac{1}{2}$ *b.* $\frac{1}{3}$ *c.* $\frac{1}{2}$ *d.* $\frac{3}{4}$

2. Why are the oceans salty?   *a.* The oceans have always been salty.   *b.* Most of the salt is the result of the weathering of rock.   *c.* Most of the salt came from outer space in meteorites.   *d.* Most of the salt came from underwater volcanoes.

3. Why are tropical regions warmer than polar regions?
   a. Tropical areas are always closer to the sun.   *b.* Polar regions lose more energy to space.   *c.* The sun is usually higher in the sky in the tropics.   *d.* Tropical areas reflect more of the light that strikes them.

4. Why is food more abundant in cold ocean water than in warmer water?   *a.* Cold water absorbs more oxygen. *b.* All fish prefer cold water.   *c.* Cold water has more energy.   *d.* Cold water is more salty.

5. How does the Coriolis effect cause ocean currents to change?   *a.* Most ocean currents curve to the left in both hemispheres.   *b.* Most ocean currents curve to the right in both hemispheres.   *c.* Ocean currents generally curve left in the Northern Hemisphere and right south of the equator.   *d.* Ocean currents generally curve right in the Northern Hemisphere and left south of the equator.

Fill In

6. The continental _____ is the part of a continent that is under water.

7. The sun and the moon combine to form _____ tides, which are higher than normal.

8. Huge waves called _____ are formed from earthquakes on the ocean floor.

9. A(an)_____ starts as a sandbar then builds well above sea level.

10. Piles of rock called _____ are constructed to decrease the erosion of a beach.

Free Response

11. Describe the main parts of the ocean floor.

12. When rocks weather, they yield a variety of substances such as clay, silt, and salt. Why does salt remain in the water while most other substances settle out?

13. What are currents and how are they affected by Earth's rotation?

14. What causes tides?

15. What does it mean to say that waves do not move water?

16. Why do waves break when they approach the shore?

Portfolio

I. Report on the effect of international treaties to protect our ocean resources.

II. Investigate the progress of commercial ventures that have exploited the physical resources of the oceans.

III. If you live near the ocean, collect data and make a graph of the tide level over a period of 12 or more hours.

IV. Make a list of nonfood materials that we get from the ocean and how they are used.

Chapter 18

Earth's History

Eᴀᴄʜ ᴅᴀʏ, scientists around the world carefully record data about events that have occurred on Earth. This collection of data describes Earth's history for thousands of years. But how do scientists know when an event occurred if it happened long before people were around to record it? The clues are in the rock layers of Earth's surface. The rock layers themselves record the events. Geologists interpret the clues by relating events on Earth today with those that occurred long ago. In this chapter, you will find out what geologists have learned—and are still trying to learn—about the history of Earth.

THE FORMATION OF EARTH

Earth's history began when it was formed, along with the rest of the planets of the solar system, about 4.6 billion years ago. The solar system is made up of our sun, the nine planets, and any other objects that revolve around the sun. You will learn about the solar system in Chapter 27.

Different explanations have been proposed to describe the formation of the solar system. However, the explanation favored by most astronomers (scientists who study the planets, moons, and other objects in space) is that the solar system began as a huge cloud of dust and other matter that originated in the explosion of a massive star. The cloud, which had been slowly spinning, began to shrink as a result of the force of gravity. As the cloud collapsed, it rotated faster and faster around its own

center. The center of the cloud became so hot that a nuclear reaction formed the sun. The matter that remained in the cloud
continued to spin around the sun. Gravity caused this matter to
collect in clumps that eventually became the planets and the
moons.

When Earth first formed, it did not have the atmosphere or
the oceans it has today. So how did they form? Scientists believe
that the molten interior of Earth became a source of magma for
volcanic eruptions. These eruptions gave off a variety of gases
including nitrogen and carbon dioxide, which became abundant
in the atmosphere, and water vapor, which condensed to form
the oceans.

Scientists are not certain how life began on Earth. Two alternative models seem most likely. One model suggests that there
was a great variety of organic compounds known as amino acids
in the early ocean of the young Earth. When joined in the right
ways, the molecules of these compounds can build proteins that
might have led to the beginnings of life. Experiments to duplicate the composition and processes of the primal oceans have
been inconclusive, although planet Earth had millions of years
to produce the first living matter. Another possibility lies with
meteorites, rocks that fall to Earth from outer space. Complex
organic compounds found in some meteorites could be indi-
cations of life on other planets and might have been responsible
for bringing the seeds of that life to Earth. Perhaps Earth's life-
forms originated elsewhere in the universe. Scientists have yet to
settle this issue.

FOSSIL RECORD

Some of the best information about the history of Earth has
come from the study of fossils. **Fossils** are the preserved remains or traces of living things. Fossils include bones and teeth
of organisms, imprints of organisms left in rocks, and even foot-
prints or burrows left by organisms.

Most ancient forms of life have left few, if any, fossils. In
order for a fossil to form, the remains of an organism must be
buried in sediment soon after the organism dies. As the sediments slowly compact and harden into rock over time, the shape

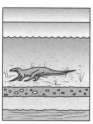 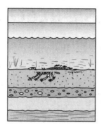  

A. An animal dies and sinks into shallow water.

B. Sediment covers the animal.

C. The sediment becomes rock, preserving parts of the animal.

D. Mountain building, weathering, and erosion eventually expose the fossil at the surface.

Figure 18-1. Fossils can form when sediment covers an animal's body.

of the organism is preserved. For this reason, fossils are usually found in sedimentary rock. They are not commonly found in igneous and metamorphic rocks because the heat and pressure associated with the formation of these types of rocks usually destroy any organic remains.

Different kinds of fossils have been found. Occasionally, organisms are preserved whole. This might happen when an organism is frozen. Ice-age mammoths, for example, were preserved in the frozen permafrost of Siberia. Other organisms were preserved whole when they become trapped in a sticky substance. Insects have been preserved in amber (tree sap), and ice-age mammals were preserved in tar pits near Los Angeles, California.

More often, only the hard parts of an organism, such as shells, bones, and teeth, are preserved as fossils after the soft parts decay. If the hard parts of an organism are replaced by minerals carried in groundwater, the organism is said to be petrified. **Petrification** means turning into stone.

Other fossils are created when an organism is pressed into soft sediment. When this happens, the shape of the organism is preserved when the sediment hardens into rock. If the soft part of the organism decays and the hard parts dissolve, an empty space is left in the rock. This empty space, which is called a **mold**, has the same shape as the organism. Minerals may then fill the mold, creating a **cast**, which is a duplicate of the shape of the original organism.

INTERPRETING THE FOSSIL RECORD

Geologists investigate events of the past by looking at fossils in relation to the bedrock in which they are found. Bedrock is rock that is a part of the solid Earth. Geologists prefer to look at bedrock because events represented in it occurred in the place where the rock now stands. Rocks that are not attached, sometimes called float, may represent events that occurred far away before the rock was transported by water or ice.

Geologists can compare the ages of different fossils by identifying the layer in which each is found. Sedimentary rocks are made of layers piled on top of each other. The **law of superposition** states that sedimentary rock layers are stacked in order of age. Each layer is older than the layer above it and younger than the layer below it. Geologists use this information to give the relative age of each rock layer. **Relative age** is the age of an object compared to the age of another object. You might describe a relative age if you say that you are younger

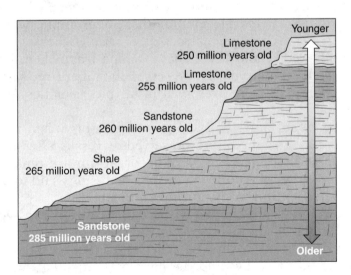

Figure 18-2. In most cases, rock layers are organized by age. The oldest layer is at the bottom and the youngest layer is at the top. The ages of fossils within the layers can be placed in order by identifying the layer in which each fossil was found.

than your mother. Relative age puts events in order of occurrence but does not give an actual number of years.

The law of superposition is based on the premise that the kinds of processes that change Earth today are similar to the processes that acted on Earth millions of years ago. This is sometimes referred to as the principle of **uniformitarianism**. For example, weathering, erosion, and deposition are actively creating and eroding sediment today. Geologists infer that these processes were also active throughout Earth's history.

What does the law of superposition indicate about fossils in the rock? The relative age of a fossil can be determined by identifying the rock layer in which it was found. For example, a fossil in a lower rock layer must be older than a fossil in a higher layer of rock.

Geologists determine the relative ages of rock layers in different locations of exposed rock, or outcrops, by using index fossils. **Index fossils** are fossils of organisms that lived during only one short period of time and are found over a wide area. The best index fossils contain parts that are readily preserved. Trilobites, an extinct group of bottom-dwelling marine animals with

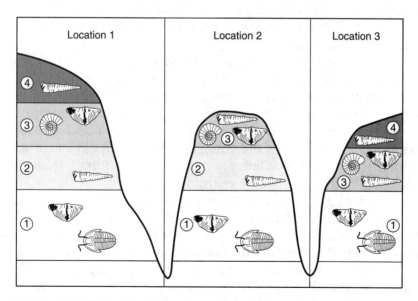

Figure 18-3. Index fossils, such as the trilobites in Layer 1, occur during only one time period but in several different locations.

a hard protective covering, are a group of animals that have been found throughout the world. Although trilobites existed for hundreds of millions of years, they changed over time. This allows for a relatively precise determination of age based on the appearance of the fossil.

CHANGES IN THE ROCK LAYERS

The law of superposition does not make it simple to determine the relative age of rock layers. This is because forces within Earth may disturb rock layers so that the layers do not show continuous deposition from the bottom to the top. Forces within Earth may lift older layers of rock up to Earth's surface. At the surface, exposed rock is then weathered and eroded. New sediment is then deposited on top of the older rock. The new sediment eventually hardens into rock as well. The place where an old eroded surface is in contact with a newer rock layer is called an **unconformity**. Since some sediment has been removed, an unconformity represents a gap in the geologic record. Geologists can study unconformities to determine when and where Earth's crust changed. In addition, by studying the fossils in these rock layers, they can find out how organisms living at the time were affected.

Faults give other clues about Earth's history. Recall from Chapter 9 that a fault is a crack in Earth's crust along which movement has occurred. Faults can occur only after rock layers have formed. Therefore, rock layers are always older than the faults they contain. The relative age of a fault can be determined from the relative age of the youngest sedimentary layer through which it cuts. Thus scientists can study the forces that have changed Earth's surface by examining faults.

Geologists can also determine the relative age of igneous rock formations. Magma often forces its way into layers of rocks. The magma hardens in the rock layers and creates a formation called an **intrusion**. Geologists can tell the relative age of intrusions because they must occur after the surrounding rock was already in place. Intrusions are always younger than the rock layers through which they pass. When magma reaches the surface and hardens, it forms igneous rock as well. Igneous

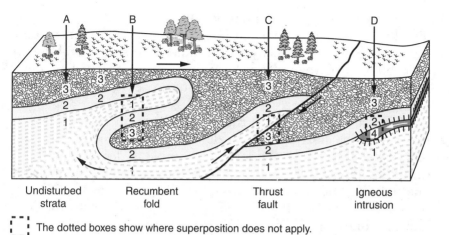

| Undisturbed strata | Recumbent fold | Thrust fault | Igneous intrusion |

The dotted boxes show where superposition does not apply.

Figure 18-4. In most places, the oldest rock layers are on the bottom, as in position A. In position B, folding has reversed the ages in the area shown. At position C, faulting has pushed older rock above younger layers. At position D, intrusion has created the youngest rock in the diagram.

rock that forms from lava on Earth's surface is called an **extrusion**. Geologists can tell the relative ages of extrusions because they are always younger than the rock layers below them.

ABSOLUTE AGE

Knowing the relative age of an event is helpful in creating a picture of Earth's history. Sometimes, however, geologists want to know the absolute age of an event or fossil. An **absolute age** gives the number of years since an event occurred or an organism lived. Scientists can determine the absolute age of rocks or fossils by using radioactive elements.

All radioactive substances are unstable. That means that their nucleus, or center of their atoms, breaks down or decays. They are said to decay from the radioactive "parent" element to a stable "daughter" element. Some elements decay into more than one element before a stable element is produced. The element into which a radioactive element decays is called a decay product.

Every radioactive element decays at a measurable and predictable rate. The rate is different for every element. Some radioactive elements decay in seconds. Others take millions or even billions of years. The decay rate is not affected by environmental factors such as temperature, pressure, or chemical changes.

The rate of decay of a radioactive element is described in half-lives. The **half-life** of an element is the time it takes for half of the radioactive element in a sample to decay. Suppose a sample contains 10 kilograms of a radioactive element. After one half-life, the remaining sample will contain 5 kilograms of the element and 5 kilograms of the decay product. In the next half-life, one half of the remaining amount, or 2.5 kilograms, will decay. For example, uranium 238 has a half-life of 4.5 billion years. In a hand-sized sample with millions of atoms, half of the atoms of uranium 238 will decay to lead 206 in 4.5 billion years. Another half of those remaining atoms will change to lead in the next 4.5 billion years. No matter how many atoms you have, half will change to the decay product every 4.5 billion years. This continues until the radioactive element is completely decayed.

Some rocks and fossils contain radioactive elements that scientists can study to determine their absolute ages. By comparing the amount of radioactive element with the amount of the decay product, scientists can determine how many half-lives have occurred. For example, if the amount of radioactive element is equal to the amount of decay element, scientists know that one half-life has occurred. So the age of the rock is equal to the half-life of the radioactive element. This procedure is referred to as **radioactive dating** or **radiometric dating**.

Uranium 238 is found in many igneous rocks. Because of its long half-life, it is useful in finding the age of very old rocks. Carbon 14 is another radioactive material. It is found in all living things. It has a relatively brief half-life—5700 years—which makes it most useful in finding the age of recent sediments or rocks that contain plant or animal remains. Rocks and fossils that are more than 50,000 years old cannot be dated using carbon 14 because all, or most, of the carbon 14 will have decayed.

Radioactive dating has allowed scientists to estimate the age of Earth. Before scientists used radioactive materials to de-

Radioactive Decay Data

RADIOACTIVE ELEMENT	DISTINTEGRATION	HALF-LIFE (years)
Carbon 14	$C^{14} \rightarrow N^{14}$	5.7×10^3
Potassium 40	$K^{40} \underset{Ca^{40}}{\overset{Ar^{40}}{\rightleftarrows}}$	1.3×10^9
Uranium 238	$U^{238} \rightarrow Pb^{206}$	4.5×10^9
Rubidium 87	$Rb^{87} \rightarrow Sr^{87}$	4.9×10^{10}

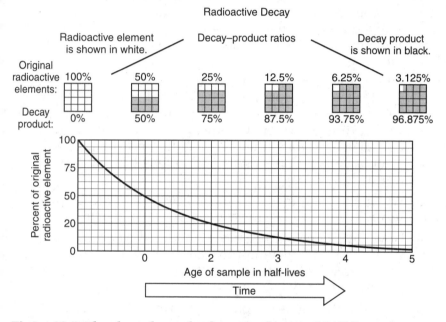

Figure 18-5. The chart shows the decay products and half-lives of several radioactive elements. The graph shows that half of the atoms change to the decay product during the first half-life. In each additional half-life that passes, half of the remaining material changes to the decay product.

termine the age of rocks, scientists could not agree on Earth's age. They used such inexact methods as estimating how much time was needed to make the oceans salty, how long it would take for the outer layers of the Earth to cool from molten rock, or how long it would take to deposit all of the layers of sedimentary rock. But now, measurements of radioactive elements in the oldest known rocks and similar testing of meteorites (bits of rock that fall to Earth from outer space) have yielded an age of 4.6 billion years.

QUESTIONS

Multiple Choice

1. What is believed to be the source of Earth's atmosphere and the oceans?   *a.* gases given off in volcanic eruptions *b.* liquid and gas that fell from outer space   *c.* a collision with another planet   *d.* gases and liquids ejected by the sun

2. Which organism would be especially useful as an index fossil?   *a.* an organism that was alive for a brief time and found over a small area   *b.* an organism that was alive for a short time but found worldwide   *c.* an organism that was alive for a long time but found over a small area *d.* an organism that was alive for a long time and found worldwide

3. Which part of a fossil organism is *least* likely to be preserved?   *a.* teeth   *b.* muscle   *c.* bones   *d.* shell

4. How have some animals been preserved whole as fossils? *a.* They become metamorphosed with the bedrock. *b.* They are covered by lava flows.   *c.* They are preserved in frozen ground.   *d.* They have melted to form magma.

5. According to the principle of uniformitarianism,   *a.* all rocks were made from molten rock   *b.* geologic processes active today were also active in the past   *c.* the oldest rocks in an area are usually found near the top   *d.* 4.6 billion years was not enough time to make Earth what it is today.

6. In an exposure of bedrock, the oldest layers are usually found   *a.* near the bottom   *b.* near the top   *c.* in the middle   *d.* spread throughout the layers.

7. How can scientists identify an unconformity in a bedrock exposure?   a. Some layers are thicker than others. *b.* Fossils are absent from the outcrop.   *c.* The older layers are on the bottom.   *d.* The rock record is not continuous.

8. Which part of a radioactive sample remains unchanged after two half-lives? *a.* one-quarter *b.* one-half *c.* three-quarters *d.* none of it

9. Which is the most reliable method for determining the age of Earth? *a.* calculating how long it took the oceans to become salty *b.* determining how long it took to deposit all of the Earth's sedimentary rock *c.* measuring radioactive materials in Earth's oldest rocks *d.* estimating how long it took Earth to cool from a molten mass of magma

10. How old is Earth? *a.* about a million years *b.* about 4.6 million years *c.* about a billion years *d.* about 4.6 billion years

Fill In

11. The preserved remains of living things are called _____.

12. A fossil in which the hard parts of the organism have been replaced by minerals carried in groundwater is said to be _____.

13. According to the law of _____, sedimentary rock layers are stacked in order of age.

14. The _____ age of a fossil compares it to the age of other fossils but does not give its age in terms of a specific number of years.

15. Radioactive dating can be used to determine the _____ age of a fossil.

Free Response

16. Summarize the events through which Earth is believed to have been formed.

17. Why are fossils found primarily in sedimentary rocks?

18. How do geologists use the law of superposition to find the relative age of fossils?

19. What is an unconformity?

20. How can scientists determine the age of a fossil by measuring the amount of a radioactive element that is present?

Portfolio

I. If you live in an area where fossils are found, make a collection of fossils. Use them to establish the age of the rocks and relate any other information they tell you about the geologic past in this area.

II. Make a diagram of a rock outcrop in your area that has many layers. On your diagram, show or otherwise indicate the observable characteristics of the different layers and which are the oldest.

III. Look for structures such as folds and faults. Record the change produced in each structure you are observing.

Chapter 19
Geologic Time

THROUGH ANALYSIS OF THE FOSSIL RECORD and radioactive dating, geologists have been able to develop a clear picture of Earth's history. That history is characterized by numerous and varied changes to Earth's surface as well as to the organisms that inhabit Earth. Geologists have also been able to put these events in order on a geologic time scale. The units of measurement on this scale are far greater than any calendar you might use. Rather than hours, days, or months, the geologic time scale must accommodate the millions of years that pass between events. In this chapter, you will learn how geologists have divided Earth's history into useful units, and you will find out about significant events during these divisions.

THE GEOLOGIC TIME SCALE

Geologic time begins with **Precambrian time**. Precambrian time covers about 87 percent of Earth's history—from the time Earth was formed about 4.6 billion years ago to 545 million years ago. The name comes from the fact that rocks containing some of the earliest fossils were found in the Cambrian region of Great Britain. (Cambria was the old Roman name for Wales.) All the earlier layers were at first thought to be void of fossils and were therefore called Precambrian. Geologists now know that there were organisms alive in Precambrian time, but their fossils can be difficult to recognize because the organisms lacked hard parts. Geologists have found fossils of single-celled organisms,

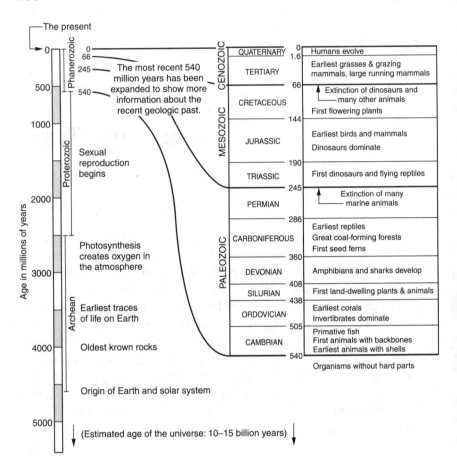

Figure 19-1. Geologic time scale.

such as algae, in rocks that formed about 3.5 billion years ago. Although scientists do not know exactly where or when life on Earth began, they do know that it occurred during this span of time because of these fossils.

After Precambrian time, the basic units of geologic time are eras and periods. The longest segments of Earth's history are **eras**. There are three eras: Paleozoic, Mesozoic, and Cenozoic. Each era is further divided into **periods**. The names of the geological time periods, such as Cambrian, Ordovician, and Silurian, were chosen according to the region in which rocks of that age were first studied.

THE PALEOZOIC ERA

The Paleozoic Era is characterized by a more abundant fossil record than Precambrian time. This is because so many new life-forms with bones and shells that fossilized appeared during this time span. The name *Paleozoic* can be translated as "the time of early life." This era is divided into six periods.

The Paleozoic Era included the development of many forms of marine life including shelled animals and fish as well as the first plants and animals to inhabit the land. According to studies of ancient fossils, many species perished at the end of the Paleozoic Era. During this mass extinction, in which many types of living things became extinct at the same time, as much as 95 percent of the life in the oceans disappeared. The only animals to survive were fishes and many reptiles. Although some species of plants became extinct, others survived.

Such an extreme episode of extinctions is not easy to explain. Some scientists have suggested that the formation of the supercontinent Pangaea caused dramatic environmental and climatic changes. However, such a catastrophic event probably was the result of a more drastic cause. Perhaps the impact of a large meteor and the resulting volcanic eruptions killed many organisms and caused a huge change in Earth's temperature.

THE MESOZOIC ERA

The Mesozoic Era began about 250 million years ago. The name for this era, which fits in the middle of the other two, means "middle life." This era is divided into three periods.

The fish, insects, reptiles, and other forms of life that survived the mass extinction of the Paleozoic Era became the main forms of life at the beginning of the Mesozoic Era, or the Triassic Period. In fact, this era is also called the age of reptiles because they were so successful during this span of time. Dinosaurs and the first mammals also appeared during the Triassic Period. The first mammals were small animals about the size of a mouse. The Triassic Period was followed by the Jurassic

Period. Dinosaurs became the dominant land animal during this period. The first birds also appeared in this period. During the final period of the Mesozoic Era, the Cretaceous Period, flowering plants evolved.

At the end of the Mesozoic Era, another mass extinction occurred. More than half of the plant and animal groups became extinct. Dinosaurs were one of the many organisms to disappear entirely. The reason for this mass extinction is also unclear. However, many scientists hypothesize that it was a result of the impact of an asteroid. (An asteroid is a rocky mass from space.) They suggest that the impact of the asteroid caused huge amounts of dust to enter the atmosphere, thereby blocking out the sunlight. The lack of sunlight caused the plants to die, which in turn caused the plant-eating animals to starve. Other scientists hypothesize that the mass extinction was a result of climatic changes resulting from increased volcanic activity. The true explanation may even involve factors not yet understood.

THE CENOZOIC ERA

The last era is the Cenozoic Era, which means "recent life." This era, which is divided into two periods, is often called the Age of Mammals because the extinction of dinosaurs created an opportunity for mammals to flourish. During this span of time, mammals adapted to life on land, in water, and in the air.

During the first part of the Cenozoic Era, the Tertiary Period, Earth's climates were generally warm and mild. However, during the Quaternary Period, Earth's climate cooled, causing the series of ice ages you read about in Chapter 16. One of the most significant events during this period is the emergence of the ancestors of modern humans.

The periods of the Cenozoic Era are further divided into **epochs**. These smaller time spans are used because the fossil record for this era is more complete than those before it. Using epochs makes it easier to organize the fossil record.

We are still in the Quaternary Period of the Cenozoic Era. The processes of change will continue, and only the passage of time will complete the time scale.

ORGANIC EVOLUTION

One of the most obvious lessons learned from studying the changes that occurred across the geologic time scale is that Earth has been home to a great diversity of organisms. Collecting fossils from various locations over the past few centuries has indicated that the first life-forms, generally those found in the lower layers, are the most simple. Moving upward in the layers, which is forward in time, we see a greater diversity of life-forms. These observations have led scientists to the conclusion that organisms have changed through time. The process by which different kinds of organisms change over long periods of time is called **evolution**. More specifically, evolution is a change in species over time. A species is a group of organisms that are similar and can interbreed to produce fertile offspring. Dogs, for example, make up a species.

One of the driving forces behind evolution is mutation. A **mutation** is a change in the material that organisms pass on to their offspring. The material consists of **genes**, which are units of heredity. Genes carry a complex code that determines the characteristics of an organism. If a gene mutates, the change will produce a difference in the organism's offspring.

In most cases, mutations cause improper development. The organism with the mutation is unable to compete with other organisms and dies off quickly. In rare cases, however, the mutation benefits the organism by making it better suited to its environment. A change that increases an organism's chance for survival is called an **adaptation**. Organisms that survive produce offspring with the same adaptation. Over a long period of time, many small adaptations may lead to the evolution of a new species that no longer resembles its ancestors.

The geologic time scale indicates that most of the fossils found today are the remains of organisms that are extinct or no longer exist on Earth. Extinction occurs when a species is no longer able to survive in its environment. This might happen if the environment changes and the species does not adapt to it. It might also happen if a new species evolves that is better adapted to the environment. The original species cannot compete successfully for food and other resources and dies out.

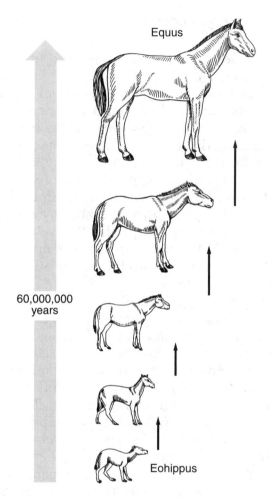

Figure 19-2. Evolution of the horse.

Evolutionary changes in the forms of organisms over time enable geologists to estimate the age of rock layers. For example, any remains of dinosaurs would establish rocks as Mesozoic. Trilobites, marine animals with a hard covering that lived in the ancient oceans, would indicate a Paleozoic age of the rock in which they are found. But both of these groups of animals changed through time. Therefore, a fossil of a particular appearance might allow a geologist to establish the specific part of the era during which the organism lived.

THE MAGNITUDE OF GEOLOGIC TIME

In considering the history of planet Earth it is easy to talk of millions of years as if we could readily imagine such long periods of time. But in the context of our lives, we might consider a week, a month, or a year to be a long time. Converting the geologic time scale to something we can appreciate may help us to understand how brief our own lives are in Earth's history. So imagine a clock that measures one day.

In the first minute after midnight, dust and other debris clump together into a hot ball of lava that we now call Earth. The Earth's outer layers solidify, and the most primitive life-forms appear around 7:00 A.M. After about 21 hours, at about 9 P.M., more complex organisms evolve in the ocean. An hour later, reptiles and insects appear. Dinosaurs rule Earth from about 11 P.M. until they become extinct at about 11:30 P.M. In the last 20 minutes or so life is dominated by mammals. Modern humans appear in the last second before midnight. In fact, all of recorded history is less than the last tenth of a second, and the entire span of a human life is only on the order of a thousandth of a second.

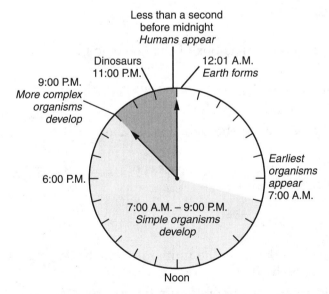

Figure 19-3. This clock places Earth's history in 24 hours. All human history occurs within the last second.

QUESTIONS

Multiple Choice

1. The time period from which the fewest fossils have been found is   *a*. Precambrian time   *b*. Cambrian Period   *c*. Mesozoic Era   *d*. Cenozoic Era.

2. According to the geologic time scale, which part of Earth's history spanned the longest period of time?   *a*. Precambrian   *b*. Paleozoic   *c*. Mesozoic   *d*. Cenozoic

3. Which kind of organism has probably existed for the longest period of time on Earth?   *a*. mammals   *b*. dinosaurs   *c*. algae   *d*. fish

4. What has happened to most of the organisms we find as fossils?   *a*. They now live in the deepest parts of the oceans.   *b*. They have become extinct.   *c*. They have moved other planets.   *d*. They come out only at night.

5. Among a variety of individual organisms, which are the most likely to survive?   *a*. the largest individuals   *b*. those that have the fewest offspring   *c*. the smallest individuals   *d*. those best suited to their environment

Fill In

6. The longest divisions of the geologic time scale are called _____.

7. The era during which dinosaurs lived was the _____.

8. The period in which we live is the _____.

9. The periods of the Cenozoic Era are further divided into _____.

10. Through the gradual process of _____, living organisms change over time.

Free Response

11. List and describe the units into which the geologic time scale is divided.

12. How has our understanding of Precambrian time changed in the past?

13. What do scientists know about the Paleozoic Era?

14. What characteristics distinguish the Mesozoic Era from the others?

15. Describe a significant event of the Cenozoic Era.

16. What is evolution?

Portfolio

I. Use a long roll of paper, like those sold for adding machines, to make a time line for planet Earth. Divide your time line into Precambrian, Paleozoic, Mesozoic, and Cenozoic according to the length of each. Show the times of several of the most important events in each.

II. Visit a nearby museum that has pictures or dioramas of your region in the geologic past. Choose a period of time and write a report about how the area has changed.

III. Obtain a geologic map of your region or your state. Find the geologic age of the bedrock where you live and research the forms of life that existed at that time.

IV. View a science fiction film about the early Earth. Make a list of things in the film that were not accurate and explain why.

Chapter 20
Characteristics of Earth's Atmosphere

WE ARE SELDOM AWARE THAT we live in an ocean of air. From our perspective, air is usually colorless and odorless. Unless we move quickly, we rarely observe the properties of air, or even become conscious of its presence. Yet air is vital to our existence. Air makes up the lower portion of Earth's atmosphere. The **atmosphere** is a thin blanket of gases that surrounds planet Earth. In this chapter, you will learn about the composition and layers of the atmosphere and how energy moves throughout the atmosphere.

COMPOSITION OF EARTH'S ATMOSPHERE

Air is a mixture of several different gases. Nitrogen (N_2) makes up 78 percent of the volume of air. Nitrogen is changed by some bacteria into a form that can be used by other organisms. However, nitrogen is relatively inert, meaning that it usually does not take part in natural chemical reactions.

Oxygen (O_2) is the second most abundant component of the atmosphere, comprising 21 percent of air's volume. It is an active element that readily reacts with other materials. Our bodies get energy from the reaction of oxygen with the food we have eaten. Oxygen is also necessary for the reactions of rusting and burning. If oxygen were more abundant in the atmosphere, fires would be more common and destructive.

Nitrogen	78%
Oxygen	21
Argon	1
Carbon dioxide	0.03
Other gases	Less than 1%

Figure 20-1. Composition of dry air in the lower atmosphere by volume. Water vapor can vary from less than 1% to as much as 3%

The other 1 percent of air consists mostly of argon (Ar) and carbon dioxide (CO_2). Helium (He), hydrogen (H_2), krypton (Kr), and ozone (O_3) are other gases present in very small amounts. **Ozone** is formed when sunlight in the upper atmosphere causes oxygen to undergo a chemical reaction. Ozone, which is concentrated in a region known as the ozone layer, is essential to life on Earth because it absorbs most of the harmful ultraviolet rays from the sun. You will learn more about the ozone layer in Chapter 23.

Air also contains water vapor (the gaseous form of water). Water enters the atmosphere by evaporation from the oceans and returns to Earth as precipitation. The water vapor content of air is highly variable. It depends on temperature, which in turn depends on such factors as the season, time of day, and location. Near the poles, for example, the air is cold and can hold very little water vapor. But the warm air in the tropics may contain as much as 3 percent water vapor. You will learn about water vapor in the air in Chapter 21.

Finally, air contains small particles such as dirt, pollen, soot from fires, and chemicals. These particles are generally classified as dust.

LAYERS OF THE ATMOSPHERE

Although the atmosphere extends more than 100 kilometers above the visible surface of Earth, most of its mass is concentrated below an altitude of 10 kilometers. Like any other matter on Earth, the atmosphere is pulled downward by the force of gravity. This gives the atmosphere weight. The weight of the air above makes the bottom of the atmosphere relatively dense. The air becomes thinner and thinner as it extends upward into space.

Above 100 kilometers the atmosphere is so thin that its properties are difficult to measure, so it has no precise upper limit.

The atmosphere is divided into layers, or shells, based on
trends in temperature. (You were introduced to the layers of the
atmosphere in Chapter 2.) The lowest layer, the **troposphere,**
extends from the surface to an altitude of approximately 10 to
15 kilometers. It is thicker in the tropics and thinner near the
poles. Within the troposphere, temperature gradually decreases
with increasing altitude. This is why snow can be found on high
mountaintops even in the summer. The world's highest mountain, Mount Everest in Asia, reaches close to the top of the troposphere. Clouds and water vapor are generally restricted to the
troposphere, which is why weather patterns are observed only in
this layer.

The second layer of the atmosphere is the **stratosphere.**
The stratosphere extends to an elevation of about 50 kilometers.
Unlike conditions in the troposphere, the temperature within

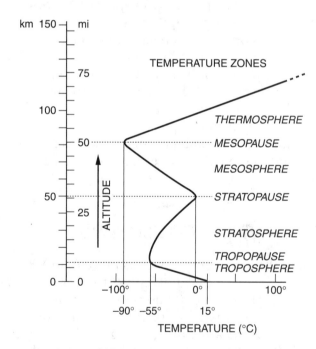

**Figure 20-2. Earth's atmosphere is divided into
layers according to changes in temperature.**

the stratosphere steadily increases with altitude. The stratosphere is clear and dry. Most of Earth's ozone is found in this layer. Jet aircraft generally fly in the stratosphere because of its strong, steady winds and good visibility.

The third layer of the atmosphere is the **mesosphere**. In this layer, temperatures fall rapidly with altitude to the height of about 80 kilometers, which is the coldest part of the atmosphere. The fourth layer of the atmosphere is the **thermosphere.** In this layer temperatures again rise as nitrogen and oxygen absorb solar energy. However, the air is so thin that air temperature has little meaning.

ENERGY

When you note conditions in the atmosphere, such as rain or wind, you are describing weather. **Weather**, which changes continuously, is the short-term condition of the atmosphere. People who study weather conditions are known as **meteorologists**.

Changes in the weather result from the movement of energy through the atmosphere. There are several different forms of energy, but in this case we are talking about thermal energy. Recall that matter is made up of individual particles that are in constant motion. These particles have energy as a result of that motion. The total energy of all the particles in a sample of matter is called **thermal energy**. Thermal energy can be transferred from one sample of matter to another. Thermal energy that moves from a substance at higher temperature to a substance at lower temperature is known as **heat**. There are three methods of heat transfer.

One method of heat transfer is **conduction**. During conduction, heat is transferred when two objects are in direct contact. Heat is transferred from a stove to a pot by conduction. Similarly, air in contact with warm ground is heated by conduction.

The second method of heat transfer is **convection**. During convection, heat is transferred by the movement of currents within a fluid. When a pot of water is put on a stove to boil, the water at the bottom of the pot is heated by conduction. The heated water expands and becomes less dense than the colder water above it. So the colder water sinks and the warmer water rises. In time, this pattern of rising and sinking forms a current that causes all of the water to become heated. The current is known as a convection current.

The third method of heat transfer is **radiation**. Objects radiate energy mainly in the form of electromagnetic waves. (Visible light, X rays, and microwaves are examples of electromagnetic waves. Electromagnetic waves differ by wavelength and frequency.) Very hot objects radiate primarily electromagnetic waves with short wavelengths, such as visible light and ultraviolet light. Cooler objects radiate electromagnetic waves with longer wavelengths, such as infrared waves. Radiation is the only form of heat flow that can pass through a vacuum, such as outer space.

The sun radiates energy into space in all directions. Energy, or radiation, from the sun is called **insolation**. This word comes from the contraction of three words—*in*coming *sol*ar radi*ation*. Relative to the other objects in the solar system, Earth is a small planet about 150 million kilometers from the sun. So Earth receives only a tiny fraction of the sun's energy. Of the energy Earth does receive, about 30 percent is reflected back into space by particles and water vapor in the atmosphere and by Earth's surface. The remaining energy is absorbed by Earth's surface or particles in the atmosphere.

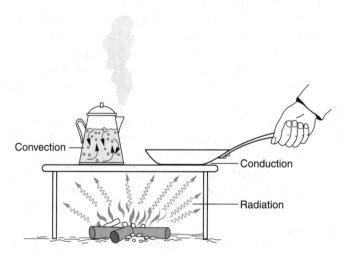

Figure 20-3. Heat can be transferred by conduction, convection, or radiation.

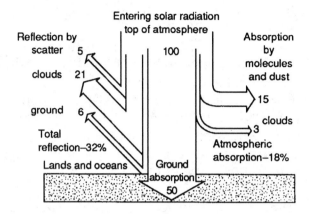

Figure 20-4. Of the insolation that arrives from the sun, half is absorbed by Earth's surface. The rest is reflected to space or absorbed by the atmosphere.

You may wonder why Earth does not heat up if it constantly absorbs energy. The answer is that Earth radiates energy as well in the form of infrared waves. In addition, through conduction, heat is transferred from the ground to the lowest layer of air. Then through convection, the heat is carried through the atmosphere.

Not all of the energy radiated from Earth escapes into space. The atmosphere traps some of it. This phenomenon is known as the **greenhouse effect**. Gases known as greenhouse gases are responsible for this effect. The infrared waves warm the atmosphere as such greenhouse gases as water vapor and carbon dioxide in the air absorb them. You will learn about the greenhouse effect in Chapter 23.

QUESTIONS

Multiple Choice

1. The most abundant gas in the atmosphere is   *a.* oxygen *b.* carbon dioxide   *c.* nitrogen   *d.* argon.

2. Ozone forms from   *a.* carbon dioxide   *b.* argon
   c. osmium   *d.* oxygen.

3. Which variable is used to separate the atmosphere into
   four layers?   *a.* temperature   *b.* humidity   *c.* wind direction   *d.* cloud types

4. The lowest layer of the atmosphere is the   *a.* stratosphere
   b. troposphere   *c.* mesosphere   *d.* thermosphere.

5. The two atmospheric layers in which temperature
   decreases with altitude are the   *a.* troposphere and mesosphere   *b.* troposphere and thermosphere   *c.* stratosphere
   and thermosphere   *d.* stratosphere and mesosphere.

6. The method of heat transfer that involves currents within
   a fluid is   *a.* radiation   *b.* conduction   *c.* insolation
   d. convection.

Fill In

7. Jet aircraft usually fly within the layer of the atmosphere
   known as the _____.

8. The layer in which most of the mass of the atmosphere
   can be found is the _____.

9. A(an) _____ is a person who studies the weather.

10. Heat transfer by _____ involves electromagnetic waves.

11. Energy radiated from the sun to Earth is also known as

    _____.

12. As a result of the _____, some infrared radiation from
    Earth is trapped in the atmosphere.

Free Response

13. Describe the composition of air.

14. Why isn't the mass of the atmosphere spread evenly
    throughout?

15. On what are the boundaries between atmospheric layers
    based?

16. Compare the three methods of heat transfer.

17. Does Earth absorb all of the isolation is receives? Explain
    your answer.

Portfolio

I. Research the composition of atmospheres of other planets to determine how they compare with Earth's atmosphere.

II. Prepare a list of questions that you would ask a meteorologist to find out what such a career entails and what training is required. With your teacher's permission, contact a local meteorologist to conduct your interview. Present your results to the class.

Chapter 21
Atmospheric Moisture

WHEN YOU LOOK OUT THE WINDOW, how do you describe the weather? You may note whether it is raining or the sun is shining. You may consider how warm or cold it feels. You may even notice how the wind is blowing. In other words, you are identifying the basic characteristics of weather. When meteorologists describe weather, they specify five variable quantities: temperature, humidity, air pressure, precipitation, and wind. In this chapter, you will learn about some of these characteristics and will find out how they are related to moisture in the atmosphere.

TEMPERATURE

You learned in Chapter 20 that heat is thermal energy transferred from one substance to another. When a substance absorbs heat, one of two things happens. Either its temperature rises or its state changes. We will consider these possibilities separately.

When a substance absorbs heat, the amount of thermal energy it possesses increases. As a result, its molecules vibrate faster. The energy associated with motion is called kinetic energy. The faster air molecules move, the more kinetic energy they have. **Temperature** is a measure of the average kinetic energy of the molecules in a sample of matter. As kinetic energy increases, temperature rises. So the more energy the molecules of air have, the hotter it feels. (Do not confuse the definition of temperature with the definition of thermal energy. Temperature is average kinetic energy. Thermal energy is total energy.)

Temperature is measured with a **thermometer.** Thermometers operate on the principle that most liquids expand when they are heated. Most thermometers consist of a liquid confined in a narrow tube with a bulb at the bottom. The liquid is often mercury, which is silver in color, or alcohol, which is usually dyed red. As the air gets warmer, the liquid expands and rises in the tube.

A temperature scale is printed on or near the tube of a thermometer. The scale shows the units of temperature, which are called **degrees.** The two basic temperature scales are the Fahrenheit and Celsius scales. The Celsius scale is used in the metric system. The scales are organized by assigning a value to certain fixed temperatures, such as the temperatures at which water freezes and boils. The difference between those two values is then divided into equal units. On the Fahrenheit scale, water freezes at 32° and boils at 212°. Since the difference between those values is 180, each Fahrenheit degree is 1/180 of the difference. On the Celsius scale, the fixed points are 0° and 100°. Therefore, each Celsius degree is 1/100 of the difference.

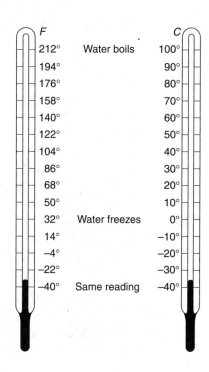

Figure 21-1. The Fahrenheit and Celsius temperature scales.

STATES OF MATTER

The more energy a substance has, the faster its particles move. At some point, the particles of matter gain enough energy to break free from their fixed positions. How free the particles of matter are to move determines the state of matter—solid, liquid, or gas. So a change in energy can cause matter to change from one state to another.

Figure 21-2 shows how water changes from a solid to a liquid and then to a gas as energy is added. The process in which water changes from a solid to a liquid is called **melting,** and the process in which liquid water changes to water vapor is called **evaporation** (or vaporization). The reverse processes are **condensation** (water vapor to liquid water) and **freezing** (liquid water to solid ice). Note that as energy is added to an ice cube, the substance does not increase in temperature at a constant rate. In addition, the temperature does not increase at all during the stages at which water is changing from one state to another.

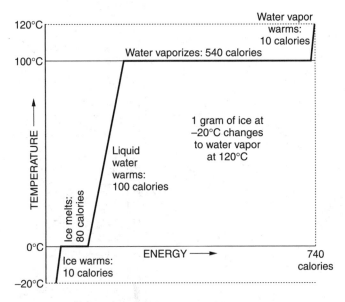

Figure 21-2. Thermal energy is measured in units called calories. For the energy-absorbing processes shown in this graph, vaporization alone accounts for nearly three-quarters (540) of the total 740 calories.

So the temperature of the solid rises until it changes into a liquid. Then the temperature of the liquid rises until it changes into a gas.

Energy involved in a change in state is known as **latent energy** because it does not change the temperature. Latent energy is absorbed during melting and vaporization. Latent energy is released during condensation and freezing.

HUMIDITY

The water vapor content of the air is called **humidity**. The amount of water vapor the air can hold varies greatly with temperature. Warm air can hold far more moisture than cold air can. Meteorologists usually express humidity as **relative humidity**, which compares the actual amount of water vapor in the air with the maximum amount of water vapor the air can hold at the given temperature. Relative humidity is expressed as a percent. If the relative humidity is 50 percent, the air contains only half as much water vapor as it could at its present temperature. If the air contains the maximum amount of water vapor (100 percent), the air is said to be saturated.

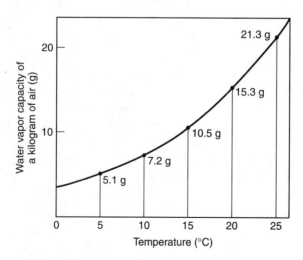

Figure 21-3. Air's capacity for water vapor increases with temperature.

Suppose during the day a sample of air has a relative humidity of 75 percent at a given temperature. At night, however, the temperature of the air drops rapidly. Since the amount of water vapor the air can hold depends on temperature, the relative humidity will increase as the temperature drops. At some point, the air will reach a temperature at which it becomes saturated and can hold no more water vapor. This temperature at which saturation occurs is called the **dew point**. Cooling beyond the dew point causes water vapor in excess of the air's capacity to come out of the air through condensation. If the water vapor condenses on surfaces such as plants or grass, it is called dew. It may also condense into droplets that make up a cloud or fog. If the temperature at the surface is below 0°C, the water vapor condenses as a solid known as frost.

CLOUDS

When air above Earth's surface cools below the dew point, clouds form. A **cloud** is a collection of very small ice crystals or water droplets in suspension in the atmosphere. Because most of the atmosphere is very cold, most clouds are made of ice crystals. Only the lowest clouds are composed of water droplets.

What might cause the temperature of air to drop and a cloud to form? Warm air near Earth's surface rises because it is less dense than the cool air above it. The air encounters less air pressure as it rises. Air pressure, which you will learn more about in the following chapter, is the gravitational force holding the atmosphere near Earth. Air pressure at a particular point is a result of the weight of the air above that point. Therefore, air pressure decreases as altitude increases. Lower air pressure means that the air molecules can move farther apart. As molecules move farther apart, they collide less often and transfer less energy. Less energy translates to lower temperature, so the air cools as it rises. This cooling effect is known as **adiabatic cooling**.

One additional condition must be met in order for clouds to form. Water vapor needs something to condense onto. Such particles are known as **condensation nuclei**. Tiny particles of dust,

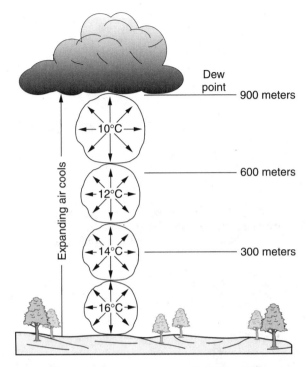

Figure 21-4. When moist air rises and cools adiabatically below the dew point, clouds form.

salt from ocean spray, and smoke from fires provide the condensation nuclei needed for cloud formation. Clouds will not form without condensation nuclei.

PRECIPITATION

As the ice crystals or water droplets float around in a cloud, they bump into one another and join together. The droplets get larger and heavier until they are heavy enough to fall from the sky as **precipitation**. It can take a million cloud droplets to make a single raindrop. Most precipitation starts as snow, which can turn to rain if it falls through warmer air before reaching the ground.

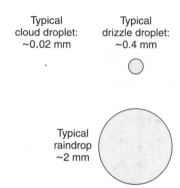

Typical
cloud droplet:
~0.02 mm

Typical
drizzle droplet:
~0.4 mm

Typical
raindrop
~2 mm

Figure 21-5. Large drops of rain form by collecting smaller droplets. This diagram shows cloud, drizzle, and raindrops magnified 10 times.

QUESTIONS

Multiple Choice

1. Freezing on the Fahrenheit scale is assigned the temperature of   *a.* 0°   *b.* 32°   *c.* 100°   *d.* 212°.

2. Boiling on the Celsius scale is assigned the temperature of   *a.* 0°   *b.* 32°   *c.* 100°   *d.* 212°.

3. The process in which liquid water changes into water vapor is called   *a.* freezing   *b.* evaporation   *c.* melting   *d.* condensation.

4. The ability of air to hold water vapor depends on   *a.* air pressure   *b.* wind speed   *c.* temperature   *d.* wind direction.

5. Which two things need to be present in air for a cloud to form?   *a.* nitrogen and oxygen   *b.* nitrogen and condensation nuclei   *c.* water vapor and oxygen   *d.* water vapor and condensation nuclei

Fill In

6. The average kinetic energy of the molecules in a sample of air is measured by its _____.

7. The temperature scale used in the metric system is the _____ scale.

8. The energy given off or absorbed during a change of state is called _____.

9. The _____ compares the amount of water vapor in air with the maximum amount it can hold at a given temperature.

10. The temperature of air drops as it rises due to _____ cooling.

11. In clouds, water vapor condenses around small particles known as _____.

Free Response

12. How does a thermometer work?

13. What is latent energy?

14. What does it mean to say that the relative humidity of air is 40 percent?

15. Why do clouds form when air reaches its dew point?

16. How is air cooled as it rises into the atmosphere?

17. What could prevent a cloud from forming even if air is cooled below the dew point?

Portfolio

I. Make a cloud in a bottle. (1) Run hot water into the bottom of the bottle. (2) Add smoke by blowing out a match inside the bottle and holding it there for a few seconds. (3) Cover the top of the bottle with a plastic bag of ice and water. CAUTION: Obtain adult assistance with this activity.

II. Find the dew point by slowly adding ice to water in a metal cup. Place a thermometer in the cup to measure the falling temperature as you watch for water condensing on the outside of the cup. Condensation begins when the water in the cup reaches the dew point temperature.

Chapter 22
Air Pressure and Winds

JUST AS CHANGES in the amount of water vapor in the atmosphere create a variety of weather conditions, so too do movements of the air. Knowing how and why air moves through the atmosphere is important to understanding weather conditions and weather prediction. In this chapter, you will learn how the movement of air masses causes changes in weather and how meteorologists use information about winds and air masses to create weather maps and make predictions about the weather.

ATMOSPHERIC PRESSURE

You may be familiar with different examples of pressure. For example, bicycle and automobile tires require specific pressures. Pressure is generally defined as a force per unit area. **Atmospheric pressure**, also called air pressure, is the weight of the atmosphere per unit area. It results from the gravitational force holding the atmosphere near Earth.

In the atmosphere, air pressure generally decreases as altitude increases because the pressure at any point is a result of the weight of the air above that point. At higher altitudes, there is less air above and therefore less weight to support. The pressure at any point is equal in all directions.

Air pressure is measured with a **barometer**. In a mercury barometer, air pressure pushes down on the surface of mercury in a dish. This causes a column of mercury to rise in a tube. The height of the column of mercury depends on air pressure. As air pressure increases, the column rises. As air pressure falls, the

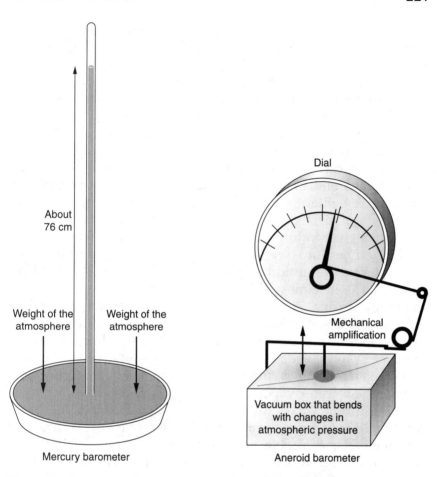

Figure 22-1. Mercury barometers use the height of a column of mercury to indicate changes in air pressure. Aneroid barometers use changes in the shape of an airtight container.

column falls as well. Air pressure readings are given according to the height of the column of mercury. The column is about 76 centimeters high at sea level, so air pressure at sea level is said to be 76 centimeters of mercury. Air pressure may also be described in the metric unit known as the millibar.

Modern barometers are generally **aneroid barometers**. This type of barometer consists of an airtight metal container that is sensitive to changes in air pressure. When air pressure increases, the walls of the container are pushed inward. When air pressure decreases, the walls of the container bulge outward. As the container changes shape, it moves a needle connected to the container. The needle is mounted on a dial that indicates the air pressure.

The National Weather Service prints weather maps that indicate the daily barometer readings. The maps show air pressure by providing actual barometer readings and by creating isobars. An **isobar** is a line that joins points with the same air pressure at a given time.

Air pressure changes as the factors on which it depends, temperature and humidity, change. In terms of temperature, warm air is lighter than cold air. Since air pressure is a result of the weight of air, the air pressure at the ground falls if warm air replaces a given volume of cold air. For the same reason, when cold air comes into a region, the pressure at the ground rises.

In terms of humidity, moist air is lighter than dry air. How can this be, since most objects, such as clothes, are heavier wet than they are dry? The answer is that the only way a gas, such as air, can absorb additional matter, such as water vapor, is if the water vapor displaces some of the air molecules. In other words, the water vapor pushes out an equal volume of dry air. Nitrogen and oxygen molecules, the primary components of air, are both heavier than water vapor molecules. (Atomic masses are measured in atomic mass units, amu. A nitrogen molecule has a mass of 28 amu, an oxygen molecule has a mass of 32 amu, and a water vapor molecule has a mass of 18 amu.) When water vapor molecules replace either nitrogen or oxygen molecules, the air gets lighter!

In general, falling air pressure indicates the approach of warm, moist air. Regions of low pressure tend to have unsettled weather conditions and precipitation. Rising air pressure usually signals the approach of cool, dry weather.

WINDS

The uneven heating of the atmosphere creates regions of high and low air pressure. Air moves from regions of higher pressure to regions of lower pressure. The resulting movement of air is **wind**. The greater the change in air pressure over a given distance, the faster the speed of the wind.

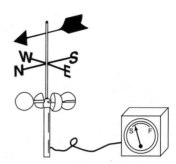

Figure 22-2. An anemometer, which measures wind speed, is often combined with a wind vane, which tells wind direction.

The movement of local winds, winds that extend over a distance of 100 kilometers or less, follows the pattern just described. However, a weather map shows that winds do not blow directly from high pressure to low pressure when it comes to large regions of high and low pressure. This is due to the Coriolis effect. Recall from Chapter 17 that ocean currents are curved due to Earth's counterclockwise rotation. Winds are curved for the same reason. In the Northern Hemisphere, winds curve to the right. In the Southern Hemisphere, winds curve to the left.

How can you tell the direction and speed of the wind? Wind direction is measured with a **wind vane**. A wind vane has a broad tail that resists the wind. If, for example, the wind is blowing from east to west, the tail will be pushed to the west. The head of the wind vane will point into the wind. Thus a wind vane always points to the direction from which the wind is blowing. For this reason, meteorologists specify wind direction as the direction from which the wind is coming rather than the direction in which it is moving. Thus an east wind is air moving from the east toward the west. The speed of wind is measured with a device called an **anemometer**.

AIR MASSES

Air is not always in motion. When a body of air stays over a location for a few days or more, it takes on the characteristics of the weather conditions of that location. If the air remains over the ocean, it becomes moist. If it stays in the Arctic, it becomes

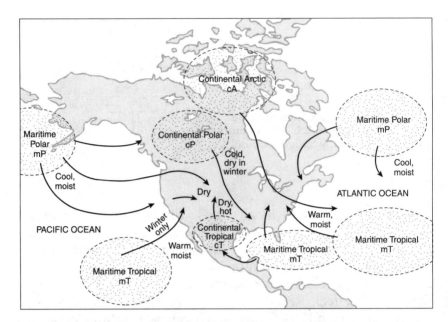

Figure 22-3. The temperature and humidity characteristics of an air mass depend on where it originated.

cold. A huge section of the lower troposphere that has the same kind of weather throughout is called an **air mass**. Air masses can be so large that two or three air masses are enough to cover the continental United States at one time.

Air masses are generally characterized according to their temperature and where they originate. So they may be either tropical (warm) or polar (cold), and either continental (dry) or maritime (moist). An air mass moving into the United States from central Canada during the winter would probably be a continental polar air mass. An air mass coming into the United States from the Gulf of Mexico would probably be a maritime tropical air mass. Meteorologists use the origin of an approaching air mass to predict the type of weather to expect.

FRONTS

The boundary between any two air masses is called a **front**. If one air mass displaces another, the passage of a front leads to a change in the weather. The extent of the change depends on the differences between the two air masses. Most precipitation comes with fronts.

There are four kinds of fronts: cold, warm, occluded, and stationary. **Cold fronts** mark the approach of cold air. If cold, dry air moves into an area with warm, moist air, the heavier cold air will wedge itself underneath the warm air. As the warm, moist air is pushed up, it expands and cools, causing cloud formation and precipitation. Cold fronts pass quickly, generally in an hour or less. Soon after a cold front moves in, the weather changes to cool and dry.

Warm fronts move more slowly than cool fronts. In a warm front, warm air is pushing ahead and displacing colder air. The lighter air behind a warm front has little weight to push the cooler air it is displacing. The warm, moist air moves up and over the air mass ahead of it, causing a gradual thickening and lowering of clouds. As a warm front passes, it commonly results in several days of warm, rainy weather.

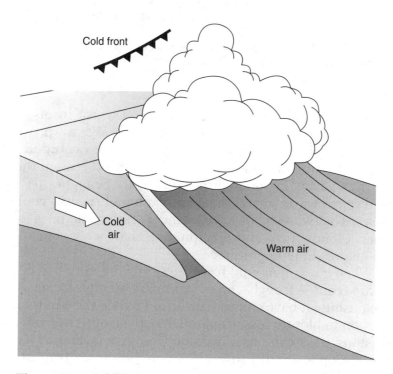

Figure 22-4. Cold fronts pass quickly as cold, dry air wedges under air that is warmer and more moist.

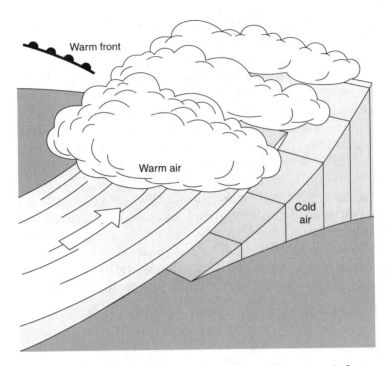

Figure 22-5. Warm fronts pass slowly because warm air has little power to push the air mass ahead of it.

Because cold fronts move faster than warm fronts, sometimes a cold front catches up with and overtakes a warm front. When a cold front overtakes a warm front, an **occluded front** results. The slow-moving warm air is trapped between the cool air it was advancing on and the cold air that caught up with it. The trapped warm air is pushed up on top of the two cooler air masses. The two cool air masses underneath sometimes join together. Clouds form and precipitation may occur, but there is little change in temperature.

When a warm or cold front stops moving, it becomes a **stationary front**. A front becomes stationary because the winds that were pushing it have changed direction. When the winds change direction to blow across, or parallel to, the front, the front slows and stops. This can happen if one front meets another coming from the opposite direction. It is common for a front to slow down and stall over a region for several days. The front then may simply disappear over the course of a few days. Stationary fronts

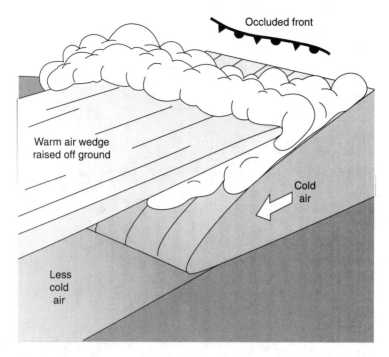

Figure 22-6. When a cold front overtakes a warm front, the warm, moist air is isolated aloft.

become moving fronts again if the winds change direction to blow perpendicular to the front. In the area of a stationary front, skies are usually cloudy. If there is enough humidity, there may be precipitation for several days.

PREDICTING WEATHER CONDITIONS

Meteorologists use three different resources to predict weather. First, they need to know the present weather conditions over a broad region. Second, they need to know the history of the weather conditions in that area. Third, they need computers to process the data.

To find out what the present weather conditions are, meteorologists obtain information from several weather stations in every state. Ships and airplanes provide information on conditions over the oceans. Weather satellites supply information for the entire Earth.

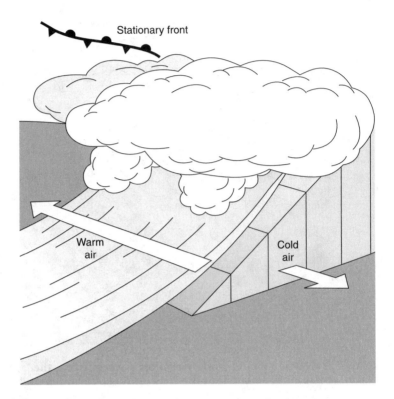

Figure 22-7. Winds blowing in opposite directions along a stationary front minimize the mixing of air masses.

By carefully analyzing past weather systems, meteorologists have learned how weather systems develop. Knowing how air masses changed and interacted in the past improves meteorologists' ability to predict how weather will change in the future.

From the constant stream of data on present weather conditions, computers can draw continuously updated weather maps that enable meteorologists to visualize atmospheric conditions. In addition, computers can be programmed to model past developments of weather systems. This allows meteorologists to see how conditions similar to the present weather have changed through time. They can then use the past patterns to predict the course of the current conditions.

Weather maps help us to understand the present weather and to see how it is changing. Weather maps can show many conditions—temperature, air pressure, wind strength and direction, cloud cover, precipitation, and frontal movements. When several of these conditions are combined on a single map, we call that map **synoptic**. *Synoptic* comes from two Greek words meaning "see together."

In Figure 22-8, a midlatitude cyclone is moving into the eastern part of the United States. This map corresponds to the early stage of the development of an occluded front. The following weather predictions are based on the information shown on map. Portland, Oregon: cool and unsettled weather is associated with the stationary front just to the south. Salt Lake City, Utah: cool and stable weather with precipitation in the nearby mountains. Little Rock, Arkansas: continued cool, clear, and dry as high pressure dominates. Knoxville, Tennessee: a rapidly approaching cold front will bring brief showers followed by cooler, clear, and dry weather. Detroit, Michigan: Overcast and showers to continue for the next day or so, followed by cooler, clear, and dry weather. New York, New York: Rain ending, followed by warmer, hazy weather.

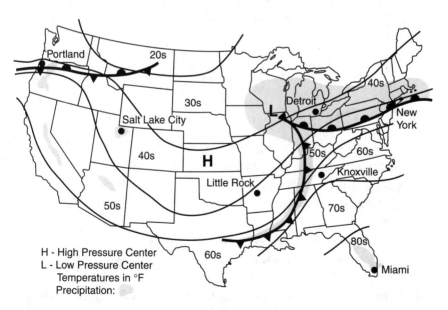

Figure 22-8. Synoptic weather maps give information about weather factors over a geographic area. This map shows the temperature, high- and low-pressure centers, and three weather fronts.

ORDER AND CHAOS IN WEATHER PREDICTION

One of the first people to program a computer to simulate weather development was Edwin Lorenz, at the Massachusetts Institute of Technology in the 1960s. The first time Lorenz ran the program, he got the expected weather conditions, but when he ran his program a second time, the results showed very different weather conditions. Lorenz figured out that the data used in the second run had been rounded off by the computer.

Although the difference between his original data and the rounded-off data was tiny, it caused very different results. Scientists call this **extreme sensitivity** to initial conditions. Our ability to successfully predict future weather conditions is dependent on how precisely we know present weather conditions. There are limitations to how well we can observe and measure present conditions. Because of these limitations, our weather predictions will be highly inaccurate if we try to predict more than a week or two into the future. Understanding the reason for the unpredictable nature of future events is an important concept of the new scientific field called **chaos theory**.

QUESTIONS

Multiple Choice

1. Air pressure is measured with *a.* a hydrometer *b.* a thermometer *c.* an anemometer *d.* a barometer.

2. Which kind of air mass is the most dense? *a.* warm and moist *b.* warm and dry *c.* cold and moist *d.* cold and dry

3. A wind that moves from the south to the north is a *a.* north wind *b.* south wind *c.* vertical wind *d.* reverse wind.

4. Which kind of air mass would probably originate in desert regions of the southwestern United States during the summer? *a.* maritime tropical *b.* maritime polar *c.* continental tropical *d.* continental polar

5. Fronts are characterized by *a.* cold weather and clear skies *b.* warm weather and stable conditions *c.* changing temperatures and cloudy weather *d.* sunny skies and steadily rising temperatures.

6. Which change is most likely as a cold front approaches? *a.* increasing air pressure *b.* increasing temperature *c.* decreasing wind speed *d.* decreasing cloud cover

Fill In

7. Winds curve as a result of the _____ effect.

8. On a weather map, a thin line called a(an) _____ connects regions with the same pressure.

9. Wind direction is measured with a _____.

10. A large body of air that is relatively uniform in temperature and humidity is called a(an) _____.

11. A place where two different air masses come together is known as a(an) _____.

12. A(an) _____ is a frontal boundary that does not move.

13. A(an) _____ map combines several weather conditions.

14. The unpredictable nature of events is related to the field known as _____ theory.

Free Response

15. Why does air pressure decrease with altitude?

16. How do mercury barometers measure air pressure? How do aneroid barometer measure air pressure?

17. Why does moist air rise?

18. What does falling air pressure indicate about weather conditions? What does rising air pressure indicate?

19. What is an air mass and how are air masses described?

20. Describe the four types of fronts.

21. How do meteorologists predict the weather?

Portfolio

   I. Use the Internet to access current weather maps. Regional radar images of clouds and precipitation are especially useful. Print out several maps and report on the weather that day.

  II. Cut daily weather maps out of the newspaper and display them in sequence to show weather development and storm movements.

 III. Make your own weather predictions one day in advance and compare your reliability with predictions from television or your newspaper.

 IV. Measure two atmospheric variables, such as temperature and air pressure, to find out how changes in one affect the other.

Chapter 23
Evolution of the Atmosphere

So FAR, Earth is the only planet known to be able to support human life. All the other planets in our solar system are either too hot or too cold. They have atmospheres that are too dense or too thin. In addition, some atmospheres would be poisonous to humans. No other planet has an atmosphere that contains enough oxygen to support human respiration.

As you learned in Chapter 18, all the planets in our solar system were formed from debris left by an exploding star. This matter came together by the force of gravity into clumps that eventually formed the planets. If the formation of Earth was typical of the formation of the sun's other planets, why do we see evidence of life only on Earth? In this chapter, you will learn more about the formation of Earth's atmosphere, how it supports life as we know it, and how human activities have altered the atmosphere over time.

THE FIRST WATER

Liquid water is essential for life, but Earth did not always have water. About 4.6 billion years ago, planet Earth was a mass of molten rock with no liquid water. Earth's first atmosphere was made up of hydrogen and helium. These are very light gases. It is believed that most of the hydrogen and helium escaped into space. Earth's second atmosphere was very rich in carbon dioxide. The carbon dioxide, along with many other gases, came from the molten rock. This process of "outgassing" is still going on today every time a volcano erupts. The carbon dioxide in the atmosphere trapped heat from both the molten rock and the sun. Because of the carbon dioxide in the atmosphere, Earth cooled very slowly.

Much of the gas that escaped from the molten rock was actually water vapor. Many scientists also believe that Earth received water vapor from the asteroids and comets that constantly bombarded Earth during its formation. While the planet slowly cooled, the water vapor condensed to form clouds and precipitation. Over millions of years the oceans filled. Eventually, the amount of water evaporating from the oceans became equal to the amount of rain that fell from the sky, just as it is today.

The distance between Earth and the sun is very important. If Earth had been farther from the sun, the atmosphere might not have been able to trap so much heat, and Earth might have cooled more. In that case, all of the water might have frozen, as it has over the North and South poles. If the Earth were closer to the sun, heat trapped by the carbon dioxide in the atmosphere would have made Earth too hot for life.

THE FIRST OXYGEN

Some scientists believe that life originated from random chemical reactions within the oceans. Other scientists say that organic materials found in some meteorites provided the beginnings of life on Earth. But they all agree that early organisms used carbon dioxide from the atmosphere.

As organisms became more complex, they developed photosynthesis. Photosynthesis is the process by which plants use the energy of sunlight, the substances in their environment, and carbon dioxide to build living tissue. During photosynthesis, oxygen is given off as a by-product. Photosynthesis provided the first atmospheric oxygen on Earth. At first, oxygen was actually toxic to many early life-forms. Over time, as the level of oxygen continued to rise, life-forms evolved that could tolerate oxygen. Eventually, animals evolved, which are totally dependent on oxygen to support their life functions.

HOW HUMANS HAVE CHANGED EARTH'S ATMOSPHERE

Our atmosphere took billions of years to develop. Yet technological advances have given today's society the power to change the atmosphere in fewer than 100 years. Unfortunately, it is not change for the better. The Industrial Revolution that began in the late eighteenth century made it possible for inventors to create many energy-saving machines. The only energy these machines save is human energy, because the machines use energy in the form of fossil fuels. Recall that fossil fuels are made of large amounts of carbon. They were created when the organic remains of plants and animals were buried millions of years ago, mostly during the Carboniferous

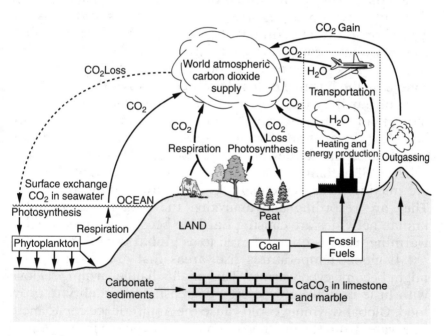

Figure 23-1. Carbon cycles naturally among the atmosphere, oceans, and land. Organic material that changes into carbonate rocks, such as limestone, and fossil fuels, stores great quantities of carbon. Humans have changed the balance of atmospheric carbon by using fossil fuels. The dotted box shows the recent human influence.

Period. Over time, heat and pressure changed the remains into coal, oil, and natural gas. When people burn fossil fuels, huge amounts of carbon dioxide and other gases are released into the atmosphere. The result is several changes to the composition of the atmosphere.

Global Warming

People who grow flowers year-round often use a special building that has walls and a roof made of glass. The transparent glass allows insolation to enter but prevents heat loss by radiation. So when the insolation enters the greenhouse, the sun's energy is trapped and warms the greenhouse. This keeps the greenhouse warmer than its surroundings.

Carbon dioxide and water vapor are like the glass in a greenhouse. They allow insolation to reach Earth's surface, but they cut down on the escape of energy. This is known as the **greenhouse effect**. Without the greenhouse effect, too much energy would escape from Earth, making it too cold. However, some scientists fear that carbon dioxide added to the atmosphere by the burning of fossil fuels could upset the energy balance of Earth by trapping more heat. This would warm Earth more than natural warming would. Furthermore, forests are being cut down at an alarming rate. Trees and other plants are important because they absorb carbon dioxide from the atmosphere. Climatologists, scientists who study different climates, are making measurements of temperatures around the world. They are recording and analyzing the data to determine if human activities are causing Earth to become warmer. Such a warming trend is often referred to as **global warming**.

If global temperatures rise, areas that are already warm might become even warmer. This might change farming regions with little rainfall into deserts, decreasing our ability to grow food. Global warming could cause the Antarctic ice cap to melt. The water from the melting ice could raise the sea level and possibly flood coastal cities.

All scientists agree that we need to monitor the effects of increasing carbon dioxide levels so that we are prepared for the possible new conditions. The future is not out of our control. If

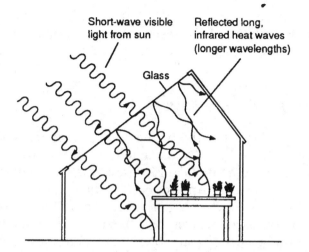

Figure 23-2. Like a garden greenhouse, Earth's atmosphere is transparent to visible radiation. However, infrared radiation is reflected back toward Earth by greenhouse gases.

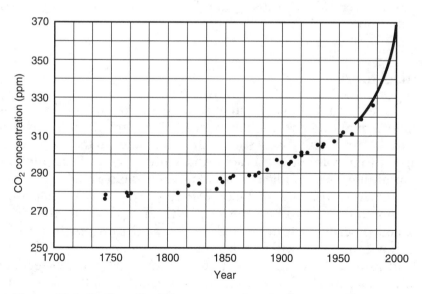

Figure 23-3. Carbon dioxide concentrations in the atmosphere.

society can decrease the use of fossil fuels, it might be possible to restore the balance of the gases in the atmosphere. As a result, we might be able to slow down excessive global warming.

Holes in the Ozone Layer

You've learned that the ozone layer is responsible for protecting Earth from harmful ultraviolet radiation from the sun. However, measurements taken in the late 1970s showed a hole in the ozone layer over Antarctica. The hole steadily grew over the next 20 years. A similar hole occurs over the Arctic, and a thinning of the ozone layer has been noted at all latitudes. Ozone thinning is the result of human activities that release gases called choloroflurocarbons (CFCs) into the atmosphere. CFCs contain chlorine, fluorine, and carbon. CFCs are used as coolants in air conditioners and in the making of foam products. When released into the atmosphere, the CFC molecules break down, and their chlorine atoms become involved in a chain of reactions that breaks down ozone in the presence of sunlight. If ozone is broken down faster than it is formed, the ozone layer thins, exposing life-forms to the sun's dangerous radiation.

Acid Rain

The burning of coal and exhaust from cars affect the atmosphere in another way—they release nitrogen and sulfur oxides into the air. These gases react with water vapor in the air to form drops that contain nitric acid and sulfuric acid. Precipitation containing these acidic drops is known as **acid rain**.

Once on Earth's surface, acid rain can be quite harmful. It can cause soil to become so acidic that plant growth is reduced or crops are destroyed. When acid rain reaches lakes, it can cause the water to become too acidic to support fish. Acid rain can also cause certain building materials, especially limestone, marble, and cement, to weather rapidly. The main way to reduce acid rain is to reduce air pollution.

QUESTIONS

Multiple Choice

1. Where did the water in the oceans come from? *a.* from the sun   *b.* from volcanic eruptions   *c.* from meteorites *d.* from the moon

2. How might Earth be different if it were farther from the sun?   *a.* Its water would be frozen.   *b.* It would be too hot for life.   *c.* It would be made of helium.   *d.* Its atmosphere would contain more carbon dioxide.

3. Which gas is needed by humans to support respiration as well as to protect us from harmful radiation?   *a.* oxygen *b.* nitrogen   *c.* carbon dioxide   *d.* water vapor

4. Which process is responsible for the abundant oxygen content in Earth's atmosphere?   *a.* combustion   *b.* meteorite impacts   *c.* photosynthesis   *d.* gravitational forces

Fill In

5. Earth's first atmosphere is believed to have been comprised of hydrogen and _____.

6. Earth cooled slowly after it formed due to the _____ in the atmosphere.

7. Photosynthesis is carried out by _____.

8. The _____ effect is the process by which carbon dioxide and water vapor trap thermal energy radiated from Earth.

9. A warming trend on Earth, known as _____, might cause water levels to rise as the Antarctic ice cap melts.

10. _____ in CFCs breaks down ozone in the atmosphere.

Free Response

11. How do current theories explain the origin of Earth's water?

12. How did the first organisms survive without oxygen in the atmosphere?

13. How can the greenhouse effect be both good and bad?

14. What is the ozone layer and what is happening to it?

15. What are the causes of acid rain? What are the potential threats?

Portfolio

I. Find out and report on laws that have been instituted in your community to reduce air pollution.

II. Investigate global warming through a variety of sources. Why is it controversial? Why is the scientific community divided on this issue?

Chapter 24

Storms

As MUCH AS WE CAN STUDY and admire the power of nature, we cannot control it. The same atmospheric changes that lead to afternoon showers and gentle breezes can also lead to devastating storms. A storm is a violent disturbance in the atmosphere. The energy of a storm is latent energy released as water vapor in rising warm air condenses into liquid water. In this process, 540 calories of energy are released for each gram of water vapor that changes to liquid water. That's enough energy to give an average hurricane about 200 times the total rate of energy production of the United States power grid! In this chapter, you will learn about the different types of storms and what you can do to protect yourself.

THUNDERSTORMS

A **thunderstorm** is a violent event caused by the upward movement of warm, moist air. Thunderstorms are accompanied by lightning and thunder, and usually rain. The rain often starts suddenly and hard. Some thunderstorms occur within a warm, moist air mass. These types of storms usually last less than an hour. Other thunderstorms occur along cold fronts where warm, moist air is pushed aloft by a wedge of cold, dry air. These types of storms are usually stronger and may last for several hours. Thunderstorms can occur in any part of the United States and at any time of the year. However, they are most common from May through September. Worldwide, about 2000 thunderstorms are active at any given time, mostly in the tropics.

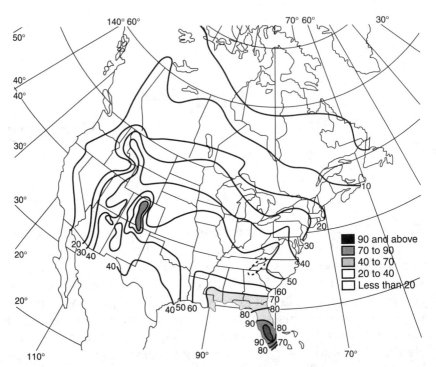

Figure 24-1. Due to the convergence of warm and moist air masses with cold air masses, Florida, the Midwest, and the Rocky Mountains get more thunderstorms than other parts of the United States.

Thunderstorms can bring several hazards, including flooding, strong winds, hail, and lightning.

Floods

Large amounts of rain can cause flooding. The danger of flash floods is especially great in arid, or dry, parts of the country. These regions do not get much precipitation overall. What precipitation they do get is in the form of infrequent but very heavy downpours.

Wind

Strong wind gusts associated with thunderstorms can blow down trees and cause a great deal of property damage. When wind topples trees onto electrical power lines, hundreds or thousands of people may be left without electricity.

Hail

Hail forms when vertical wind currents, also called updrafts, suspend raindrops at an elevation cold enough for the rain to turn to ice. The little balls of ice begin to fall through the cloud and are coated with rain. Strong vertical wind currents push the water-covered balls of ice up to a colder elevation. Here the water freezes, making the ball of ice larger. This process of falling and being covered with water, then being pushed up to a colder elevation and having another layer frozen is repeated many times. When the ball of ice becomes too heavy to be pushed up by the vertical wind currents, it falls out of the cloud and to the ground as hail. Although most hail is less than a centimeter in diameter, in the central United States hail can sometimes be as large as 7.5 centimeters in diameter. Hail can cause tremendous damage to crops and property.

Lightning

Lightning is the most dangerous feature of a thunderstorm. It is caused by a sudden discharge of electricity between positively and negatively charged regions within a cloud, between two different clouds, or between a cloud and the ground. Lightning is a more powerful example of the type of shock you might receive when you touch a doorknob after walking across a carpet. The temperature inside a lightning flash is so high that it causes the air to expand explosively. This expansion makes a loud sound that is heard as thunder.

The safest place to wait out a thunderstorm is in an interior room, away from windows, electrical wires, and water pipes. If you are caught outside when a thunderstorm strikes, keep low and crouch close to the ground so that you are the lowest object in the area. Because the metal body of a car is conductive, a car is a relatively safe place to be in a thunderstorm. The metal frame of the car will carry the charge away from you.

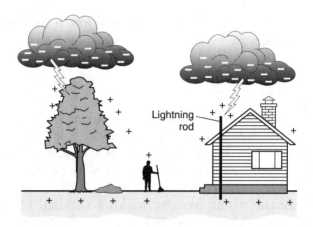

Figure 24-2. Lightning is a discharge of static electricity.

TORNADOES

A **tornado** is a narrow, funnel-shaped column of spiraling winds. The conditions that lead to strong thunderstorms can also lead to tornadoes. In fact, many tornadoes originate in thunderstorms. The center of a thunderstorm is a low-pressure area. Air inside a thunderstorm swirls in a counterclockwise direction around the low-pressure center. The swirling air, called a circulation cell, sometimes becomes powerful enough to become a tornado. A tornado that occurs over water is called a waterspout.

Although scientists do not understand the formation of tornadoes exactly, they can use radar to find, track, and measure tornadoes. Winds at the center of a tornado can range from less than 100 kilometers per hour to faster than 450 kilometers per hour. Tornadoes are commonly 150 to 600 meters in diameter.

Although they may have winds much faster than those found in the most severe hurricanes, a tornado is relatively compact and short-lived. Most tornadoes are over in less than 10 minutes, but a few have been observed to last for several hours. Tornadoes occur more often in the United States than anywhere else in the world. Within the United States, tornadoes can occur anywhere but are most common in the south-central area of the Mississippi River valley and the Great Plains.

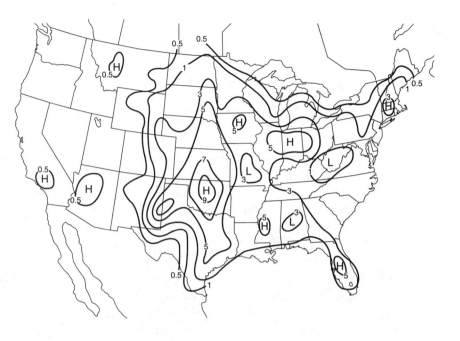

Figure 24-3. This map shows tornado frequency. Compare it with Figure 24-1, which shows thunderstorm frequency.

CYCLONES

The term *cyclone* is often mistakenly used in place of the term *tornado*. However, the two terms are not interchangeable. A **cyclone** is a wind pattern that results from an area of low pressure that contains rising warm air. As the warm air rises, more air moves in to take its place. The air currents begin to spin. Winds spiral around the center of the cyclone. In the Northern Hemisphere, the winds move in a counterclockwise direction. Cyclones are not always violent storms. The term can be used to describe anything from a slight wind to a tornado. The type of weather caused by a cyclone is often rainy and stormy.

An **anticyclone** exists in a high-pressure area that contains cold, dry air. Air descends as it spirals around and out from the center of an anticyclone. Unlike a cyclone, the winds of an anticyclone move in a clockwise direction in the Northern Hemisphere. The type of weather caused by an anticyclone is usually clear and dry.

HURRICANES

A **hurricane** is a large and powerful cyclone. Hurricanes usually begin as areas of low pressure, or depressions, over tropical waters. As warm, moist air rises rapidly, more air moves into its place. The air begins to spin. The air pressure in the center drops, drawing more air into the spinning storm. A circular wall of winds, clouds, and rainfall is formed. In the center of the storm, known as the eye, the air is calm. Outside the center, however, winds reach speeds of over 120 kilometers per hour.

Hurricanes begin during late summer and early autumn months when tropical waters are at their warmest. When the winds exceed 119 kilometers per hour, the storm is classified as a hurricane. On average, about five North Atlantic storms per year grow to hurricane status. Although the highest wind speeds in hurricanes are only about half those in tornadoes, hurricanes are more destructive because of their size, which averages between 300 and 600 kilometers in diameter. Similar storms in the western Pacific are known as typhoons, while they are called tropical cyclones in the Indian Ocean.

There are three principal sources of damage from hurricanes: wind, rain, and storm surges. Winds blow away loose objects, rip off parts of buildings, and topple trees. Rain leads to flooding. Low-lying inland areas are especially prone to flooding from the torrential rains that hurricanes bring. But an even greater danger can be the storm surge. A **storm surge** is caused by a combination of high tide, heavy rain, and high wind that builds up ocean water several meters above its normal level. Basically, the hurricane piles up water along the shore and then blows it inland. If the height of the storm occurs at high tide and in a low-lying coastal area, the combination of high water and storm waves is often deadly.

Like thunderstorms and tornadoes, hurricanes are tracked by the National Weather Service. A tropical depression first be-

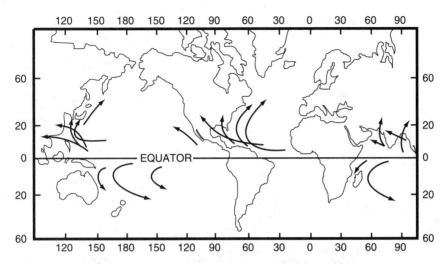

Figure 24-4. This map shows the common paths of tropical storms. These storms are known as hurricanes, typhoons, and cyclones depending on where they are located in the world. They generally begin over waters near the equator.

comes a tropical storm. When a tropical storm reaches hurricane strength, its intensity is rated by the **Saffir-Simpson scale**. This scale was developed in the early 1970s by Herbert Saffir and Dr. Robert Simpson. The Saffir-Simpson Hurricane Scale is a 1–5 rating based on the hurricane's intensity.

If you have the misfortune to be near the beach in the center of the path of a category 3 hurricane, you will notice thickening clouds and possibly a few light showers. This will be about 12

Table 24-1.   The Saffir-Simpson Scale of Hurricane Strength

Category	Sustained Winds (km/h)	Surge (m)	Damage
1	119–153	1–2	Scattered
2	154–177	2–3	Moderate
3	178–210	3–4	Extensive
4	211–250	4–6	Extreme
5	250+	6+	Catastrophic

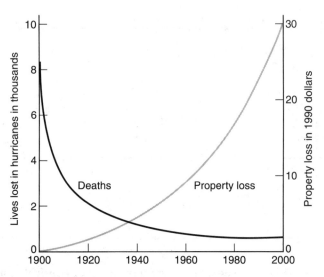

Figure 24-5. The number of deaths and injuries per year has decreased due to advanced hurricane warning and evacuation plans. But property damage has increased because of the popularity of building in exposed beachfront areas.

hours before the arrival of the center of the storm. Over the next few hours, winds will increase and rain will become more steady. A couple of hours before the center of the storm reaches you, winds will be at hurricane strength as rainfall becomes heavy. Electrical and telephone lines will be torn down. If you are very unlucky, the storm will hit at high tide with a storm surge of 3–4 meters. If you are in a low-lying area, you might experience flooding. If you are in a building facing the ocean, wave damage could be extensive. With luck, you are uninjured when the strongest winds near the center of the storm, called the eye wall, quickly die away. You are now in the calm center known as the eye of the hurricane. If it is during the day, you might notice warm temperatures and large patches of blue sky. Within a half hour you will find yourself back in the high winds and torrential rains of the eye wall. Riding out the second half of the storm might take another few hours until the sustained winds drop below hurricane strength. If you are lucky enough to survive un-

Table 24-2. Deadliest U.S. Hurricanes

Location	Year	Loss of Life
Galveston, TX	1900	6000
Louisiana	1893	2000
Central Florida	1928	1836
South Carolina/Georgia	1892	1500
Florida/Texas	1919	about 800

hurt, you will see that many of the buildings around you have been severely damaged.

WINTER STORMS

Not all storms involve warm air. When the moisture is sufficient and the temperatures are cold, the resulting storm might produce snow. These types of storms are most common in winter along the East Coast of the United States and in spring over the continental United States. A snowstorm with high winds and low temperatures is called a blizzard.

Winter storms form along the polar front. The polar front forms when cold air flows from the poles toward the equator. The cold air collides with warm air coming from near the equator. When this happens, the warm air flows on top of the cold air. The surface between the warm and cold air is called the polar front. Areas of low pressure develop along the polar front. The greater the contrast between the two air masses, the stronger the polar front and the lower the pressure.

Table 24-3. Costliest U.S. Hurricanes

Name	Location	Year	Damage ($)
Andrew	Florida/Louisiana	1992	25,000,000
Hugo	South Carolina/North Carolina	1989	7,000,000
Betsy	Florida/Louisiana	1965	6,500,000
Agnes	Midatlantic States	1972	6,400,000
Camille	Mississippi/Louisiana	1969	5,200,000

QUESTIONS

Multiple Choice

1. Lightning is caused by   *a.* the explosion of thunder   *b.* a sudden flash flood   *c.* the discharge of electricity   *d.* the eye of a storm.

2. Hail may form when   *a.* snowballs freeze   *b.* updrafts suspend precipitation as it is coated with ice   *c.* ice from the poles is carried by the polar front   *d.* a storm surge carries ice from mountaintops to lower elevations.

3. If you are caught outside during a thunderstorm, you should   *a.* lean against a tree   *b.* stand straight and still   *c.* crouch close to the ground   *d.* open an umbrella.

4. The Saffir-Simpson scale is used to describe   *a.* tornadoes   *b.* floods   *c.* hurricanes   *d.* blizzards.

5. A storm that may result from a strong polar front is a   *a.* hurricane   *b.* tornado   *c.* water spout   *d.* blizzard.

Fill In

6. Balls of ice known as _____ sometimes fall during a thunderstorm.

7. The type of storm that can produce the fastest winds is a(an) _____.

8. The highest category of hurricane on the Saffir-Simpson scale is _____.

9. A storm becomes a hurricane when its winds exceed _____ km/h.

10. The two stages that a storm goes through before becoming a hurricane are tropical depression and _____.

Free Response

11. How is latent energy related to storms?
12. How does hail form?
13. What is lightning?
14. How does a tornado form?

15. Why does a tornado usually cause less damage than a hurricane?

Portfolio

I. Prepare a report on the effects of a single historic hurricane or tornado.

II. Obtain location data for a single hurricane and plot its movement on map.

III. Use your library or the Internet to obtain data showing the number of hurricanes and tornadoes in each month of the year. Create a graph to show these data.

Chapter

25

Patterns of Climate

THINK ABOUT THE WEATHER CONDITIONS in your area over an extended period of time. Do you notice general trends? If so, you can describe the climate in your area. In this chapter, you will learn about the characteristics of climate, the three climate zones of Earth, and the factors that influence climate.

WHAT IS CLIMATE?

Climate describes the average weather conditions over an extended period of time. Every place on Earth has a specific climate. The most important elements of the climate are temperature and precipitation. Different combinations of temperature and precipitation determine Earth's major climates. For example, the climate of Florida is hot and moist, whereas the climate of the southwestern United States is hot and dry.

Several factors determine the temperature and precipitation in a region: latitude, elevation, prevailing winds, nearby bodies of water, and mountain ranges.

Latitude

Recall that latitude is the angle north or south of the equator as measured from the center of Earth. Areas close to the equator receive the direct rays of the sun. So areas near the equator have warm climates. As latitude increases, both north and south of the equator, the average temperature decreases and climates become cooler. The lowest temperatures generally occur near the poles where the sun's rays are least direct.

Elevation

Many high mountains are often capped by snow, even if they are at low latitudes close to the equator. The reason is that temperature decreases with increasing elevation. In Chapter 21 you learned that when air rises, it expands and becomes cooler (adiabatic cooling). High-altitude climates are therefore cooler than climates in nearby lowland areas.

Prevailing Wind

A **prevailing wind** blows more often from one direction than from any other direction. The amount of precipitation in a region is determined by the amount of moisture carried by the prevailing winds in that area. Areas in the path of a prevailing wind that begins over water usually receive a lot of precipitation.

Bodies of Water

Prevailing winds can also affect temperatures in coastal climates, which are along the ocean. The reason is that water has a high specific heat. **Specific heat** is the amount of energy a substance needs to absorb in order for its temperature to rise. For example, the specific heat of water is much higher than that of steel. If a mass of water and an equal mass of steel were exposed to the same amount of energy from the sun, the temperature of the steel would increase by a much greater amount than that of the water. In other words, water requires a lot more energy to raise its temperature. Because of the high specific heat of water, the temperature of ocean water does not vary much from summer to winter. The water is warmer than the air in the winter and cooler than the air in the summer. In winter, the water warms the air that moves over it. In summer, the water cools the air above it.

The winds over the United States tend to move from west to east. So during the winter on the West Coast, warm air over

Table 25-1. Specific Heats of
Common Materials

Material	Specific Heat $(\frac{cal}{g} \cdot °C)$
Water (solid)	0.5
Water (liquid)	1.0
Water (gas)	0.5
Dry air	0.24
Basalt	0.20
Granite	0.19
Iron	0.11
Copper	0.09
Lead	0.03

the Pacific Ocean blows onto land. In the summer, cooler air over the ocean blows onto land. As a result, San Francisco is warmer in winter and cooler in the summer than Washington, D.C., which is at about the same latitude. In addition, the climate along the East Coast has large seasonal variations in temperature.

Ocean currents also affect temperature. Since ocean water warms or cools the air above it, the climate of a land area near an ocean depends on the temperature of the ocean. A warm-water current will cause nearby land to have warm temperatures. Cold-water currents cause nearby land areas to have cool temperatures.

Mountain Ranges

Precipitation is also affected by the presence of mountain ranges. A mountain range acts as a barrier to prevailing winds. As air hits the mountain and moves upward, it cools adiabatically below the dew point. The excess moisture condenses and falls as precipitation. Thus the windward side of the mountain, the side facing the wind, has a wet climate. The leeward side of the mountain, the side facing away from the wind, has a dry climate. The reason is that by the time the air reaches the top of the mountain, it has lost most of its moisture. As a result,

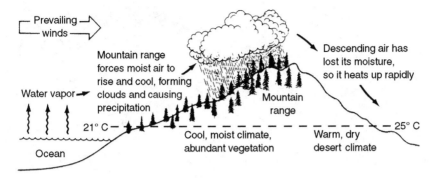

Figure 25-1. The climate varies on opposite sides of a mountain range. The windward side is wet whereas the leeward side is dry.

there is very little precipitation on the leeward side of the mountain.

CLIMATE ZONES

Climates can be divided into three general climate zones based on their temperatures: tropical, temperate, and polar. The **tropical zones** extend from the equator to about 30° north and south. Tropical climates are warm and have high humidity. There is no winter in the tropical zones, and precipitation is generally adequate throughout the year. The **temperate zones** are located between 30° and 60° latitude. Temperate climates include large seasonal changes in temperature. Although the average amount of precipitation is about the same throughout the temperature zone, some areas have heavy snow whereas areas closer to the equator have rain year-round. Temperate climates cover a huge portion of Earth. In fact, most of the United States is in the temperate zone. The **polar zones** extend from 60° to 90° latitude. Polar climates are usually very cold, with the average temperature near or below 0°C. Summers are cool in the polar zones, and there is very little precipitation.

The three major climate zones can be further classified as marine or continental (sometimes called terrestrial). Marine climates occur in areas near oceans or other large bodies of water.

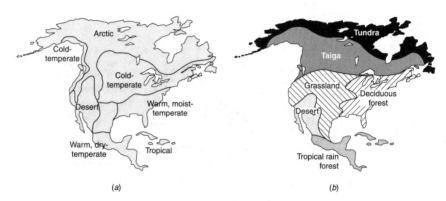

Figure 25-2. The map on the left (a) shows climate zones by patterns of temperature and rainfall. The map on the right (b) classifies climates by their plant communities. Notice that many of the boundaries between climate zones are the same on both maps.

Terrestrial climates occur within large landmasses. Marine climates receive more precipitation than do terrestrial climates. In addition, the temperatures within marine climates are more stable than those in terrestrial climates.

There are other ways to classify climates. Because living organisms are sensitive to climate, communities of plant and animal life (biomes) are often used to classify climates.

INSOLATION AND CLIMATE

You learned earlier that climate is partly a function of latitude. The reason is that the temperature in a region is largely a result of heating by the sun. Recall that sunlight and the other forms of radiant energy that Earth receives from the sun are known as insolation. In the tropics, where the noon sun is always high in the sky, the sun's rays are direct and concentrated at the surface of Earth. Therefore, the tropical zones remain warm throughout the year. In the temperate zone, summers are warmer than winters because the summer sun is higher in the sky. The polar zones are cold because they go for as long as six months without sunlight. Even when the sun is visible, it is always at an angle, which makes it appear low in the sky. So the

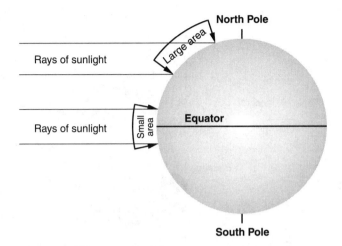

Figure 25-3. Due to differences in the angle of insolation, sunlight is more concentrated in the tropics than it is in the polar latitudes.

strength of the sunlight on the surface in the polar zones is weaker than it is in the tropics.

THE SEASONS

Locations outside the tropics have seasonal climates, warmer in the summer and cooler in the winter. The seasons are caused by the tilt of Earth's axis. Recall that the axis is an imaginary line through the center of Earth about which Earth rotates every 24 hours. Earth's axis is not straight up and down. Instead, it is slightly tilted. As Earth moves around the sun, the axis tilts toward the sun for part of the year and away from the sun for part of the year. The Northern Hemisphere has summer during the time when that part of Earth is tilted toward the sun. During that same time, the Southern Hemisphere is pointed away from the sun and has winter. At two times during the year, spring and autumn, neither hemisphere is pointed toward the sun.

The direction in which Earth points affects the insolation it receives. The hemisphere that is tilted toward the sun receives more direct sunlight than the hemisphere pointed away from the

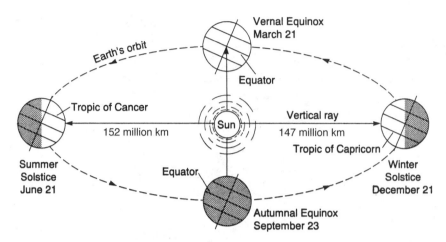

Figure 25-4. Each hemisphere experiences summer when it tilts toward the sun. The hemisphere tilted away from the sun experiences winter.

sun. The direct rays are more concentrated and result in higher temperatures.

WATER BUDGETS

A **water budget** is a numerical model of temperature and moisture availability for a particular location. Climatologists calculate the amount of water that could evaporate from the soil and plants if sufficient moisture is available. That value is called the **potential evapotranspiration**. This is a combination of evaporation from open water and from the ground with transpiration, which is the moisture given off by plants. Water budgets are based on average values of precipitation and potential evapotranspiration measured over a long period of time. They are often presented as graphs that show the relationship between the availability of moisture in soil and the moisture needed by most plants in a typical year.

A water budget that has a large deficit (shortage of water) characterizes a desert climate. A moist climate will have little, if any, moisture deficit during the year. Comparing the annual precipitation with the potential for evapotranspiration provides another way to classify climates.

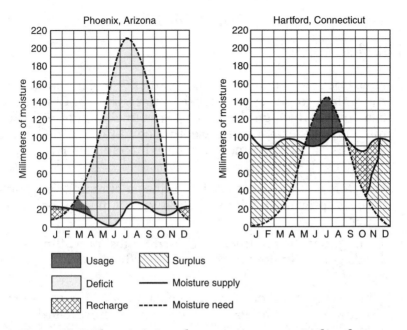

Figure 25-5. Phoenix is in a desert environment. It therefore has a large water deficit. Hartford has good precipitation throughout the year, so it has no deficit, even in the summer months.

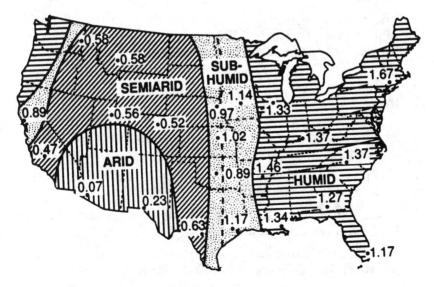

Figure 25-6. Climates can be classified by the ratio of precipitation to potential evapotranspiration.

QUESTIONS

Multiple Choice

1. How does climate differ from weather? *a*. Climate varies from place to place; weather does not vary. *b*. Climates can change through time; weather is unchanging. *c*. Climate is long-term average weather conditions. *d*. Climate includes data that could not be used to characterize weather.

2. Which is *not* a factor that determines a climate? *a*. longitude *b*. prevailing winds *c*. latitude *d*. nearby bodies of water

3. Throughout the continental United States, prevailing winds generally flow from the *a*. north *b*. south *c*. east *d*. west.

4. Which side of a mountain range usually has the most precipitation? *a*. windward *b*. leeward *c*. northward *d*. southward

5. The climate zone with the warmest temperatures is the *a*. polar zone *b*. temperate zone *c*. mild zone *d*. tropical zone.

6. The seasons result from *a*. the gradual shifting of the climate zones *b*. Earth's tilted axis *c*. Earth's rotation on its axis *d*. shifts in prevailing winds.

Fill In

7. A(an) _____ wind blows from one direction more than any other.

8. The three climate zones are the tropical zone, the _____ zone, and the polar zone.

9. _____ climates exist near bodies of water, whereas terrestrial climates exist inland.

10. The _____ Hemisphere experiences winter when the Northern Hemisphere is pointed toward the sun.

11. A numerical model of temperature and moisture availability for a particular location is called a(an) _____.

Free Response

12. Describe how the factors of latitude, elevation, prevailing winds, and mountain ranges affect the climate in a particular region

13. How does the specific heat of water result in different climates in San Francisco and Washington, D.C.?

14. What are the characteristics of the three major climate zones?

15. How does a marine climate differ from a continental climate?

16. How are the seasons related to the insolation Earth receives?

Portfolio

I. Find a variety of books that show ways to classify the climate where you live. Make a list of descriptions of the local climate from these sources.

II. Use a blank map of the United States to show the various climate zones. List the most important factors that determine the climate of each region.

III. Compute your own water budget with precipitation and potential evapotranspiration data for a city near you.

Chapter 26
Stars and the Universe

IF YOU HAVE EVER gazed up into the nighttime sky, you have undoubtedly admired the light show put on by hundreds of thousands of stars. People have studied the stars since ancient times. Stars provide light, inspiration, and a wealth of information. In this chapter, you will read about the characteristics of stars, the life cycles of stars, how astronomers study stars, and what has been learned by studying stars.

HOW FAR ARE THE STARS?

At times it seems as if stars are just out of reach. In reality, they are extremely distant. The closest star to Earth is the sun, and it is about 150 million kilometers away. That is nearly 4000 times the circumference of Earth. The next nearest star, Alpha Centauri, is about 40 trillion kilometers away. Distances such as these are too large to describe in common units. Instead, astronomers use a unit called a light-year. A **light-year** is the distance that light can travel in one year. Do not think that a light-year is a unit of time because of the word *year*. A light-year is a unit of distance. Light travels at a speed of about 300,000 km/s. (Nothing is known to travel faster than the speed of light.) At that rate, light can travel about 9.5 trillion kilometers in one year! That distance is one light-year. In terms of light-years, Alpha Centauri is about 4.2 light-years from Earth.

WHAT IS A STAR?

Stars are made up mostly of hydrogen and helium. Stars have so much mass that the conditions within their interiors make hydrogen, which is the lightest element, fuse into helium. During this process of nuclear fusion, matter is converted into energy. A tiny amount of mass results in the release of a tremendous amount of energy. This energy is slowly transferred to the visible surface of a star by convection. From there the energy radiates into space as electromagnetic energy (visible light and related forms of radiation). This is why stars shine.

THE BRIGHTNESS OF STARS

When you look into the sky, some stars appear to be brighter than others. The brightness of a star is called its magnitude. There are two kinds of magnitude: apparent magnitude and absolute magnitude. The **apparent magnitude** is the brightness as it appears to an observer on Earth. The scale of apparent magnitude was developed by ancient astronomers who classified the brightest stars as first magnitude. Stars that appeared to be half as bright were designated second magnitude, and so forth. Today each step of increase in magnitude is a difference of 2.5 times brightness. Under ideal conditions, the human eye can see stars down to a magnitude of +6. Stars that are brighter than first-magnitude stars are assigned values lower than 1. Sirius, which is the brightest of the night stars, has a magnitude of about −1.

Even though Sirius appears to be the brightest star in the sky, it is not. Sirius appears brighter only because it is closer to Earth than most other stars. Think of it as viewing a lightbulb up close and then viewing the same lightbulb from across the street. The closer you are, the brighter it looks. So how can you compare the brightness of stars that are at all different distances from Earth? Astronomers developed a scale based on the true, or absolute, magnitude of a star. The **absolute magnitude** is the inherent brightness of stars expressed as if the stars were observed at a standard distance. That distance is 32.6 light-years from the sun. At that distance, our sun would prove to be an average star with an absolute magnitude of 4.8. This means that if the sun were at a typical distance of the brightest stars we see in the night sky, it would be barely visible.

HOW HOT ARE STARS?

The surface temperature of a star is indicated by its color. A star radiates all colors of light. However, hot stars radiate more blue than red. So hotter stars appear blue and cooler stars appear red. Average stars fall somewhere in the middle, giving off light that appears more yellow. The sun, which is yellow, has a surface temperature of about 5500°C. A blue star may have a surface temperature closer to 30,000°C, whereas the surface temperature of a red star may be closer to 3000°C.

THE HERTZSPRUNG-RUSSELL DIAGRAM

In the early 1900s, astronomers Ejnar Hertzsprung (Danish) and Henry Norris Russell (American) independently plotted the surface temperatures and absolute magnitudes of stars. The resulting graph, which became known as the Hertzsprung-Russell diagram, shows that the stars formed distinct groups. The arrangement of these groups gave insight into the life cycle of stars.

Stars are born when gravity draws together dust and gases into a very large mass. Energy released in this process may create the temperatures needed to start nuclear fusion. When this happens, the energy released balances the inward pull of gravity, and the star stops contracting. Then the star quickly brightens to a stable position in the main sequence of the Hertzsprung-Russell diagram. The main sequence is the area from the upper left to the lower right. Stars within this area are called **main-sequence stars**. Ninety percent of stars we can observe fall in this region of the diagram.

A star resides in the main sequence until its hydrogen is depleted. Helium builds in the core to the point at which the energy released during fusion no longer balances the force of gravity. This process can take millions, even billions, of years.

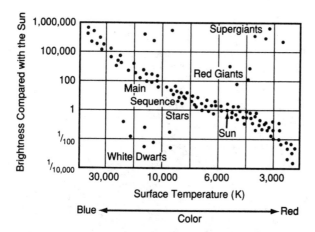

Figure 26-1. The Hertzsprung-Russell diagram relates the brightness of a star to its surface temperature.

Then the star becomes unstable, and its center starts to contract again. The center gets so hot that the outer surface expands and begins to support fusion. The star expands so much that it becomes a **red giant** or **supergiant**. These stars are indicated in the upper right of the diagram. The sun is still in its stable stage, but it too will become a red giant in about five billion years.

When a star no longer has enough fuel to supply the fusion reaction, it collapses. The star stops producing energy and gives off only its remaining energy. The result is a dim, white star known as a **white dwarf**. These stars are located toward the bottom of the diagram. A white dwarf will continue to cool until it becomes cold and dark. This process may take more than a billion years. White dwarfs occasionally become temporarily brighter again. Such a star is called a **nova**. Within a few years, novas fade back to the dim brightness of a white dwarf.

Stars that are much larger than the sun (have at least seven times its mass) can experience explosions that blow away up to half the mass of the star. During this explosion, the star becomes intensely bright and is called a **supernova**. Supernovas are so bright that for a period of time a single star may outshine all other stars in its galaxy.

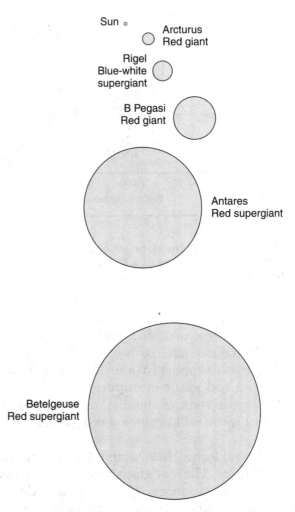

Figure 26-2. Our sun is tiny compared with giants and supergiants.

GALAXIES

Systems containing millions or even billions of stars are called **galaxies**. Astronomers have discovered that there several billion galaxies in the universe. The universe is all of space and everything in it. Most of the galaxies are millions of light-years apart from one another. That means that most of the universe is empty space.

There are three main types of galaxies. **Spiral galaxies** have a central, lens-shaped region consisting of millions of stars. Extending out from the center are spiral arms that trail behind the galaxy as it rotates. The sun is part of the Milky Way galaxy, which is a spiral galaxy. The Milky Way is so large that light takes 100,000 years to travel from one side to the other.

Elliptical galaxies can be nearly spherical or flatter and lens-shaped. Most of the stars are close to the center. New stars cannot form in elliptical galaxies because they do not contain the gas and dust required to form stars.

Irregular galaxies do not have specific shapes because the stars are spread out unevenly. They tend to be smaller, fainter, and less common than the other types of galaxies.

STUDYING STARS

How have astronomers learned so much about stars when they are so far away? The primary tool of astronomers is the telescope. The earliest telescopes were invented in the 1500s. These early devices were constructed of two lenses in a long hollow tube. Such a telescope, which uses lenses or mirrors to gather and focus light, is called an **optical telescope**. There are two main types of optical telescopes: refracting and reflecting.

A **refracting telescope** has two lenses. The objective, or primary, lens, at the front of the tube, gathers light from stars and bends (refracts) it to form an image at the rear end of the tube. The eyepiece lens at the rear end of the tube then magnifies the image for the viewer. The largest refracting telescope, which is in Wisconsin, has a primary lens about 1 meter in diameter.

Refracting telescopes are no longer built because they have several limitations. One is that a modern refracting telescope would need many lenses to correct for the fact that blue light is bent more than red light in glass lenses. Another reason is that the glass used to refract light must be perfect. Such perfection is both difficult and expensive. A third reason is that the lens can be supported only at the edges. This allows a large lens to sag over time, causing the image to blur.

The largest telescopes in use today are **reflecting telescopes**. They use a curved mirror to gather and magnify light.

Figure 26-3. This photograph shows three spiral galaxies (center) and an elliptical galaxy (far right).

For large telescopes, this design is easier to produce because the glass of a mirror does not have to be as perfect as a lens. In addition, mirrors require little color correction. A supporting framework behind the entire mirror can be used to prevent sagging. Currently, the largest reflecting telescope is in Hawaii and has a primary mirror about 12 meters in diameter.

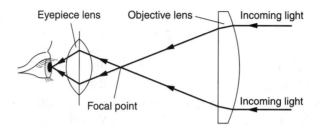

Figure 26-4. A refracting telescope uses convex lenses to form an image.

Much like the lens in a refracting telescope, the large mirror in a reflecting telescope serves as the objective. It gathers and focuses light from stars. However, in this case, the mirror is set at the bottom of the telescope tube. It reflects a small bright image back to the top of the tube where it is reflected to the observer by a small mirror. The viewer still uses an eyepiece lens to magnify the image.

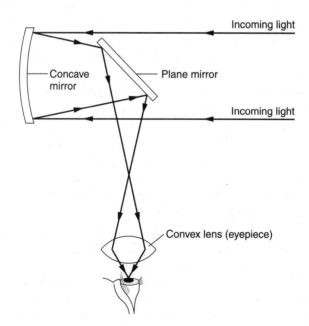

Figure 26-5. A reflecting telescope uses mirrors to reflect light.

When an astronomer selects a telescope, you might think that the decision would depend on magnification. However, magnification is not a critical issue because telescopes can be made to increase magnification by adding a different eyepiece. The magnification cannot be increased by too much, however, because as an image is enlarged, it becomes dimmer and less clear. The critical issues are the quality of the optics, especially the primary lens or mirror, and the light-gathering power. The ability to gather light from stars depends on the diameter of the primary mirror or lens. Seeing distant objects in the night sky clearly requires a balance between the largest possible light-gathering surface and the highest precision of that surface.

As fast as light travels, its speed is still limited. So when astronomers use optical telescopes to view distant objects in space, they are actually looking back in time. By the time light travels the vast distance from the star to Earth, millions of years may have passed. In fact, by the time astronomers first see light from a new star, the star may have passed through its entire life cycle. So even with the most powerful telescopes, astronomers are seeing the most distant galaxies as they were billions of years ago, nearly back at the origin of the universe.

Optical telescopes study visible light from stars, but much of the energy emitted by stars is not in the visible spectrum. Some types of telescopes detect other forms of energy. **Radio telescopes**, for example, study radio waves from space. Radio waves have longer wavelengths and lower frequencies than do visible light waves. Radio waves have been detected from the sun and other stars, from some planets, and from other galaxies. Studying radio waves has certain advantages. Radio waves can penetrate clouds and dust without being changed. In addition, radio telescopes can be used during the day whereas optical telescopes can be used at night only.

Radio telescopes can be made much larger than optical instruments because long radio waves require less precision in the reflecting surface. The largest radio telescope, located in Arecibo, Puerto Rico, is more than 300 meters in diameter. A radio telescope looks like a huge dish. The dish collects radio waves and sends them to a receiver, which converts the radio waves into electrical signals that can be analyzed. Since large radio telescopes are difficult and expensive to build, several

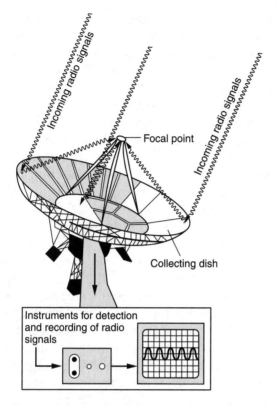

Focal point

Collecting dish

Instruments for detection and recording of radio signals

Figure 26-6. Radio telescopes use a dish to collect radiation.

smaller telescopes are often used instead. They can be arranged in arrays so that the data are combined to provide greater accuracy. Although the precise source of the radio waves can be difficult to locate, radio telescopes give us a unique view of the universe and information unavailable with optical telescopes. Instruments to detect the short wavelengths of electromagnetic energy, such as X rays and gamma rays, are also used. However, these must be located in space because such wavelengths are absorbed by Earth's atmosphere.

Another device used to study stars is called a spectroscope. Most starlight is a mixture of wavelengths. This instrument separates light into its colors. Recall that white light is actually made up of all the colors of a rainbow, or spectrum. Separating starlight into its component colors enables astronomers to discover the chemical composition of the star. The science of examining the component colors of starlight is called **spectroscopy**. The gaseous form of each element has its own unique spectrum of colors, or spectral line signature. Using spectral observations, astronomers can also infer a star's mass, size, and distance from Earth, the strength of its gravitational pull, and the composition of gases in space between Earth and the star.

Colors also provide another important piece of information—the direction in which the star is moving. This can be determined when astronomers compare the spectrum from the star with the spectrum from the same element in the laboratory. If the spectrum of the star is slightly off from that in the laboratory, it is said to be shifted. It can shift either toward the red end of the spectrum or toward the blue end. A red shift indicates that the distance between the star and Earth is increasing. A blue shift indicates that the distance is decreasing. The principle is the same as the changing sound of a train as it approaches and passes you. As the train approaches (distance decreases), the frequency of the sound waves you hear becomes higher so the sound of the train has a higher pitch. As it passes you, the sound waves spread out. The decreased frequency results in a lower pitch. This phenomenon is called the **Doppler effect**.

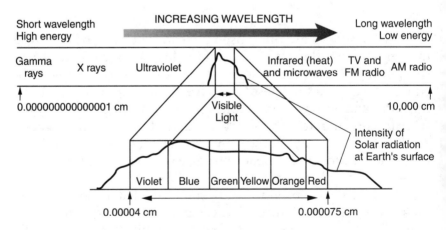

Figure 26-7. Visible light is a small portion of the electromagnetic spectrum. White light can be split into colors by a prism, which bends each wavelength by a different amount.

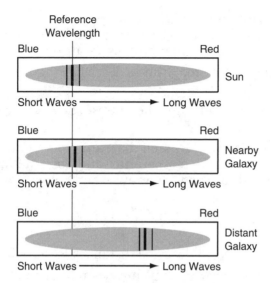

Figure 26-8. The vertical lines represent the colors of light received from a star. Notice that the lines are shifted toward the red end of the spectrum for a star in a distant galaxy. This means that the galaxy is moving away from the observer.

Red shifts discovered by astronomer Edwin Hubble in 1929 revealed that distant galaxies were receding from Earth faster than nearby galaxies. From this he concluded that the universe is expanding. If this motion is projected backward into the past, we arrive at a time when all of the mass of the universe was concentrated into a primordial object that exploded to become the universe. Such an explosion is generally called the big bang. Astronomers estimate how fast the universe is expanding to figure out how long it has been expanding. It is estimated that the universe has been expanding for 8 billion to 20 billion years.

QUESTIONS

Multiple Choice

1. The main components of stars are   *a.* hydrogen and oxygen   *b.* hydrogen and helium   *c.* iron and nickel   *d.* sulfur and iron.

2. The brightness of a star as it appears to an observer on Earth is the star's   *a*. true magnitude   *b*. actual magnitude   *c*. apparent magnitude   *d*. absolute magnitude.

3. Which absolute magnitude describes the brightest star?   *a*. 2   *b*. 6   *c*. 1   *d*. −2

4. Which color indicates the hottest star?   *a*. blue   *b*. yellow   *c*. red   *d*. green

5. The Hertzsprung-Russell diagram plots   *a*. time and distance   *b*. surface temperature and absolute magnitude   *c*. size and location   *d*. gravity and color.

Fill In

6. The unit used to describe distances to stars is the _____.

7. In stars, nuclear _____ converts matter to energy.

8. To find the _____ magnitude of a star, determine the brightness as if all stars were at a standard distance.

9. Ninety percent of the stars we can observe are _____ stars on the Hertzsprung-Russell diagram.

10. A red giant will eventually shrink into a white _____.

11. A(an) _____ such as the Milky Way is a system containing billions of stars.

12. An optical telescope that uses lenses to bend light is a(an) _____ telescope.

Free Response

13. What process occurs within stars that makes them different from other objects in the universe?

14. How does absolute magnitude differ from apparent magnitude?

15. What is the significance of the Hertzsprung-Russell diagram?

16. Are novas and supernovas the same thing? Explain.

17. How do optical telescopes differ from radio telescopes?

18. What led Hubble to the conclusion that the universe is expanding?

Portfolio

   I. Make your own telescope using a magnifying mirror and a lens.

  II. Prepare a report on the different kinds of stars.

 III. Use a map to locate the five largest telescopes. Explain why these locations were chosen.

 IV. Observe and record the colors of stars in the night sky.

  V. Investigate nuclear fusion as a potential source of energy on Earth.

Chapter

The Solar System

OF THE BILLIONS OF STARS in the sky, one makes it possible for life to exist on Earth—the sun. As you learned in Chapter 26, the sun is an average star compared with others in the universe. But the sun is the center of our solar system. The solar system contains nine planets, numerous moons, asteroids, meteoroids, and comets. In this chapter, you will learn about the solar system and the many objects within it.

THE SYSTEM OF THE SUN

All the objects in the solar system travel in paths, called orbits, around the sun or a planet. An **orbit** is the path in which one object moves around another object in space. All orbits are ellipses. Some orbits, such as Earth's, are nearly circular. However, the orbits of comets are very elongated, or eccentric.

Although we understand the organization of the solar system now, it was not always obvious to observers on Earth. This was especially true before people had telescopes with which to observe celestial objects. Until around 1600, people believed that Earth was the center of the universe. As they watched the sun appear to rise in the east and set in the west, they reasoned that the sun and the stars moved around Earth. This system of the universe is called a **geocentric**, or Earth-centered, system. Like other Greek astronomers in the second century A.D., Ptolemy believed in the geocentric system. He used the system to describe the motion of the planets as small circles moving in larger circles.

It took 1400 years for Ptolemy's model to be challenged. It was then that the Polish astronomer Copernicus proposed the idea that the sun is the center of the solar system. This is known as the **heliocentric**, or sun-centered, system. He reasoned that the planets all traveled in orbits around the sun in the same direction. The heliocentric system could be used to provide more simple explanations for observations regarding the motion of planets. Despite Copernicus's explanation, most people continued to believe in the geocentric system until about seventy years later when Galileo was able to use a telescope to observe evidence for the heliocentric system.

Not long after Copernicus, a Danish astronomer named Tycho Brahe made precise observations of the stars and planets. (He did so before the telescope was invented.) His data were passed on to a mathematician named Johannes Kepler. Kepler used the data to formulate **three laws of planetary motion.**

(1) Orbits are ellipses rather than circles. An ellipse is more oval in shape. So the distance between a planet and the sun changes. At some parts of the orbit the planet is closer to the sun, and at other parts it is farther away.

(2) The radius between a planet and the sun sweeps out equal areas in equal times. When a planet is closer to the sun, it moves faster than when it is farther away.

(3) The farther a planet is from the sun, the longer it takes to orbit the sun.

One of the consequences of Kepler's third law is that the inner planets pass the outer planets. As they do so, the outer planets seem temporarily to reverse their motion through the fixed stars. This reverse motion is known as **retrograde motion**.

Sir Isaac Newton showed that the force holding the planets (and all other revolving objects) in their orbits is the same force

Figure 27-1. The satellite orbits the sun in an elliptical path. According to Kepler's law of equal areas, the areas of A and B are equal.

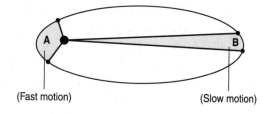

(Fast motion)   (Slow motion)

holding us on the Earth—gravity. The gravitational attraction between two objects is directly proportional to the masses of the objects and inversely proportional to the square of the distance of separation. This is known as the **universal law of gravitation**. A planet is held in orbit by the gravitational attraction between it and the sun. Without gravity, the planets would fly off in straight lines into deep space.

ROTATION AND REVOLUTION

In addition to traveling around the sun, each planet spins on its axis. Recall that an axis is an imaginary line through the center of the planet. The time it takes for a planet to complete one rotation on its axis is called its **period of rotation**. The period of rotation determines the length of one day on that planet. Earth takes about 24 hours to rotate on its axis. Thus a day on Earth is about 24 hours long. Mercury takes 59 Earth days to rotate on its axis. So a day on Mercury is much longer than an Earth day.

The time it takes for a planet to make one complete orbit, or revolution, around the sun is called its **period of revolution**. The period of revolution is equal to one year on that planet. Earth takes about 365 days to revolve around the sun. One year on Earth is 365 days. Mercury revolves around the sun in 88 Earth days. So one year on Mercury is 88 Earth days long.

THE INNER PLANETS

The planets can be divided into two groups, the inner and outer planets. The four planets closest to the sun are the **inner planets**. These planets are Mercury, Venus, Earth, and Mars. The inner planets have orbits that are all less than a third of the distance to the next planet, Jupiter. The inner planets all have a rocky crust and a dense core. For this reason they are also called the **terrestrial**, or Earthlike, planets.

The closest planet to the sun is **Mercury.** Mercury is smaller than Earth and has a smaller gravitational pull. Because of this, and its nearness to the sun, Mercury has almost no atmosphere and therefore no weather. Much of Mercury's surface is covered by craters. These craters probably formed when huge rocks crashed into the planet. The craters do not wear away over time because there is no weather to change them. Mercury's slow rate of rotation, combined with a lack of an atmosphere, causes the surface temperature to heat up to more than 400°C during the day and to cool to nearly $-200°C$ at night.

Venus is sometimes called Earth's twin because its diameter, mass, and gravity are similar to those of Earth. For many years, people wondered if Venus also had oceans and forests as well, but its thick cloud cover prevented inspection. Radar images eventually showed that the surface of Venus did have some similarities to Earth. For example, although it has a little less topographic relief than Earth, much of its surface has been shaped by volcanic eruptions. But when weather instruments were released into Venus's atmosphere, scientists discovered that the atmosphere is mostly carbon dioxide, which traps in heat like a greenhouse. This causes surface temperatures to reach close to 500°C. Researchers also found out that the thick, yellow clouds consist of droplets of sulfuric acid. And the atmospheric pressure at the surface of Venus is 90 times the sea-level pressure on Earth.

The next planet is Earth. **Earth** is the only planet on which water exists in large quantities in all three states: ice, liquid water, and water vapor. In fact, about 70 percent of Earth's surface is covered by an ocean that averages about 4 kilometers in depth. Surface temperatures range from a record low of $-89°C$ in Antarctica to 58°C in the Sahara Desert. Although the atmosphere was once mostly carbon dioxide, photosynthesis and the evolution of life have reduced the carbon dioxide content of Earth's atmosphere to a small fraction of a percent (0.03 percent). Most of the primordial carbon from carbon dioxide is now in storage in fossil fuels and carbonate rocks (such as limestone). Carbon dioxide was absorbed by plants during photosynthesis and left behind by both plants and animals to become a part of our organic sedimentary rocks. This process not only depleted the carbon dioxide content of our atmosphere but also enriched our atmosphere with oxygen. Earth is the innermost planet with a moon, and Earth's moon is relatively large compared to Earth. You will learn more about Earth and its moon in Chapter 28.

Mars is the planet that comes closest to the conditions favorable to life found on Earth. Like Venus, the atmosphere of

Mars consists mostly of carbon dioxide. But the atmospheric pressure is about 150 times less than on Earth. Mars has seasons like Earth, and temperatures along its equator can warm up to a comfortable 20°C, but the nights cool to about −60°C and the polar regions are always below about −120°C. In fact, it gets so cold that dry ice (solid carbon dioxide) forms at the poles of Mars. A canyon system has been observed on the surface of Mars. Although it looks as if it was carved out by moving water, there is no water on Mars today. This has led to some speculation that Mars was once warmer. If that water still exists on Mars, it is probably in the form of ice under the planet's visible surface. Mars also has the highest known mountain in the solar system. One of several extinct volcanoes, Olympus Mons rises to a height of about four times Earth's Mount Everest. Mars has two moons, but they are relatively small.

ASTEROIDS, COMETS, AND METEOROIDS

The inner and outer planets are separated by a ring of debris known as the **asteroid belt**. Asteroids are solid masses that are generally irregular in shape. They orbit the sun in the same direction as the planets. Most asteroids are found within the asteroid belt. However, some have a slightly oval orbit outside the main asteroid belt that brings them closer to some of the planets.

Scientists do not know for certain the origin of asteroids. They may be left over from when the solar system formed, they may be debris from a former planet, or they may be extinct comets. A **comet** is an icy object with a highly eccentric (flattened) orbit that is occasionally visible when it passes near Earth and the sun. (Comet Hale-Bopp, which passed near the sun in 1997, was the brightest comet in well over a hundred years.)

Meteoroids are also found in the solar system. A **meteoroid** is a piece of rock or icy fragment moving through space. When a meteoroid passes through Earth's atmosphere, it experiences friction. That friction heats the object and results in a streak of light known as a **meteor**. Millions of tiny meteoroids enter the atmosphere every day, yet most are burned up as they streak through the sky. Occasionally a piece of the meteoroid survives and falls to the ground. This piece is called a **meteorite**. When a meteorite strikes the surface, its impact can create a crater.

There are hundreds of known impact craters on Earth. One of the best-known craters is Meteor Crater in Arizona, which is thought to have formed 25,000 years ago.

THE OUTER PLANETS

The remaining five planets are the **outer planets**. These are Jupiter, Saturn, Uranus, Neptune, and Pluto. Of these, the first four are called **Jovian** planets, which means "like Jupiter." These planets are much larger, more gaseous, and less dense than the terrestrial planets. Pluto is not large enough to be Jovian or dense enough to be terrestrial. You will learn about Pluto at the end of this section.

The Jovian planets consist of a small, rocky core surrounded by a liquid mantle and a gaseous shell. Rather than heavy elements such as iron, silicon, and oxygen, the Jovian planets consist mainly of the light elements hydrogen and helium. The Jovian planets are also different in that they have systems of rings surrounding their equators. The rings consist of many particles that orbit the planet independently.

The first of the outer planets is **Jupiter**. Jupiter is the largest planet in the solar system. It alone accounts for two-thirds of the total planetary mass of the solar system. Hydrogen and helium make up over 99 percent of Jupiter's mass. If Jupiter were much larger, it would be a star. Jupiter rotates in just 10 hours, which is faster than any other planet. This rotation gives Jupiter a noticeable bulge at its equator. Although the pressure at the center of Jupiter is about 10 times the pressure within Earth, it is still insufficient to support the kind of nuclear fusion that powers the sun and other stars. One of the most noticeable features of Jupiter is the Great Red Spot on its surface. Some astronomers believe this to be a persistent and raging storm. Others suggest it is a calm area within a turbulent atmosphere. Jupiter has 16 moons, four of which, known as the Galilean satellites, can be seen from Earth with binoculars or a small telescope. They were named after their discoverer, Galileo.

The rings of **Saturn** make it one of the most beautiful objects in the night sky. Although all the gas giants have rings, only Saturn's are readily visible. Second in size only to Jupiter, it has a similar composition and structure with an average density less than the density of water. Saturn takes almost 30 Earth years to orbit the sun. Saturn has at least 20 moons.

Uranus was unknown to ancient astronomers because as it is not easily observed from Earth. It was not discovered until 1781. Because Uranus is so far from the sun, light is much dimmer and the surface temperature of the planet is only about −200°C. In addition, Uranus takes 84 Earth years to orbit the sun. Uranus is unusual in that its axis is tipped over. So when compared with the other planets, it spins on its side. It rotates every 17.2 Earth hours. Uranus is also unusual because the surface temperatures on the side facing the sun are about the same as those on the side facing away from the sun.

Neptune was discovered in 1846. It takes 165 Earth years to orbit the sun and 16.1 Earth hours to rotate on its axis. Its average surface temperature is −214°C. Neptune has eight moons and several rings. Its diameter is about four times that of Earth. Neptune's axis is tilted about 30° from perpendicular to its path. This causes uneven heating, resulting in seasons.

Pluto is the most distant known planet. It was discovered in 1930. Pluto is smaller and colder than any other planet, with surface temperatures predicted to be below −220°C. Pluto is more like the inner planets in composition, having a visible solid surface. It is believed to be composed of water, ice, and rocks. It takes

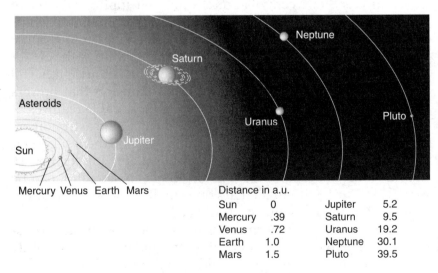

Distance in a.u.			
Sun	0	Jupiter	5.2
Mercury	.39	Saturn	9.5
Venus	.72	Uranus	19.2
Earth	1.0	Neptune	30.1
Mars	1.5	Pluto	39.5

Figure 27-2. The relative positions of the planets.

Table 27-1. The Planets

Planet	Diameter (km)	Average Distance from Sun (km)	Number of Moons
Mercury	4,878	58,000,000	0
Venus	12,104	108,000,000	0
Earth	12,756	150,000,000	1
Mars	6,794	228,000,000	2
Jupiter	142,800	778,000,000	17
Saturn	120,540	1,427,000,000	20
Uranus	51,200	2,871,000,000	17
Neptune	49,500	4,497,000,000	8
Pluto	2,200	5,913,000,000	1

Pluto 248 Earth years to orbit the sun. Unlike the other planets, Pluto's orbit is very eccentric and slightly inclined relative to the orbits of the other planets. It has a single moon, named Charon.

QUESTIONS

Multiple Choice

1. The scientist who proposed three laws of planetary motion was   *a.* Newton   *b.* Ptolemy   *c.* Copernicus   *d.* Kepler.

Table 27-2. Rotation and Revolution

Planet	Period of Rotation (Earth-days)	Period of Revolution (Earth-years)
Mercury	59	0.24
Venus	243	0.62
Earth	1	1
Mars	1.03	1.9
Jupiter	0.41	12
Saturn	0.43	29
Uranus	0.72	84
Neptune	0.67	165
Pluto	6.4	249

2. Which is a Jovian planet? *a.* Mars *b.* Saturn *c.* Earth
   d. Venus

3. The surface temperature of a planet is primarily a result of
   the planet's *a.* distance from the sun *b.* mass *c.* period
   of rotation *d.* density.

4. Which planet moves the fastest in its orbit? *a.* Mercury
   b. Earth *c.* Jupiter *d.* Pluto

5. What feature on the surface of Mars caused scientists to
   speculate that water was once found there? *a.* snow
   banks *b.* ice-covered lakes *c.* mountains *d.* channels

6. Which object is probably composed of ice and dust?
   a. Mars *b.* meteorites *c.* comets *d.* asteroids

7. If Jupiter has a composition similar to the sun, why is it
   not a star? *a.* It is in orbit around the Sun. *b.* It is too
   massive to be a star. *c.* It is too far from Earth. *d.* The
   pressure in Jupiter is too low.

Fill In

8. The path a planet takes around the sun is known as a(an)
   _____.

9. The _____ system places the sun at the center of the solar
   system.

10. The inner planets are also known as _____ planets.

11. A(an) _____ is a piece of a meteoroid that reaches Earth's
    surface.

12. _____ is the largest planet in the solar system.

Free Response

13. Compare the geocentric concept of the solar system with
    the heliocentric system.

14. Summarize Kepler's three laws of planetary motion.

15. Is retrograde motion really backward motion?

16. How is the law of universal gravitation related to the solar
    system?

17. Explain days and years in terms of rotation and
    revolution.

18. Compare the inner and outer planets.

19. Why does Venus have higher surface temperatures than Mercury, which is closer to the sun?

Portfolio

I. Consult a newspaper or the Internet to find out which planets are currently visible from your location. Then observe the planets in the night sky. Chart their motion over a one-month period.

II. Prepare a factual report about a planet in the solar system. Then develop an oral presentation about a fictitious trip to your chosen planet. Feel free to include humor in your presentation based on the conditions on the planet.

Chapter 28
Earth and Its Moon

THROUGHOUT THIS COURSE, you have learned a great deal about Earth, its characteristics, its features, and its place within the solar system. Now it is time to revisit a few concepts. In this chapter, you will take another look at time zones, rotation and revolution, seasons, and stars. Then you will consider the unique properties of Earth's moon. Finally, you will consider the significance of Earth, understanding how to take care of Earth and speculating on Earth's future as an object in the universe.

TIME ZONES

You were introduced to time zones earlier in order to understand longitude. Now that you have learned about Earth's motion, you can understand why time zones are necessary. If you could travel around Earth in a few minutes, you would see that half of Earth is in sunlight, and half is in a shadow. That means that at any time, it is daylight over exactly half of Earth and night over the other half.

For any location, we can define noon as the time when the sun is the highest in the sky and midnight as the halfway point between sunset and sunrise. If we followed this rule exactly in setting our clocks, any travel east or west would mean that we would need to change our clocks. This would be true even if traveling only a short distance, but that would be highly impractical. So a more efficient method of time change was established. The circumference of Earth is a circle, and a circle consists of 360°. That 360° is divided into 24 time zones so that each time zone differs from the next by one hour. The result of dividing 360° by 24 hours yields 15° per hour. Therefore, Earth is divided into 24 hourly time zones. The time is the same everywhere within one particular time zone, even though the apparent movement of the sun might be as much as half an hour ahead or behind in a specific region. Within most of the United States, there are four time zones: Eastern, Central, Mountain, and Pacific. (Hawaii and Alaska have their own time zones.) Although it can be a little peculiar for someone who lives near a boundary and goes to school or work on the other side, the system works quite well.

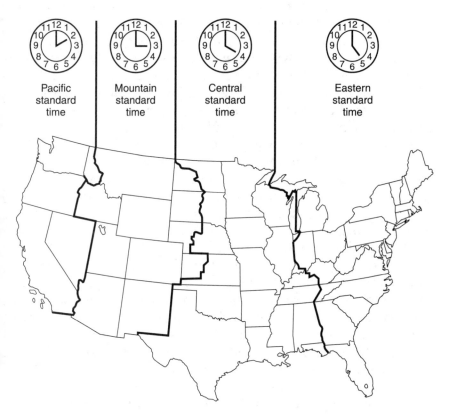

Figure 28-1. Time zones for the continental United States.

If you travel all around Earth, you will gain or lose an entire day. The **international date line** establishes the point at which the date changes. This is an imaginary line through the Pacific Ocean. If you travel west across the line, the date advances by one day. If you travel east, the date moves back one day.

EVIDENCE FOR ROTATION AND REVOLUTION

You learned that the motions of Earth were not recognized for many centuries. So what evidence is there that Earth does indeed rotate? Experimental evidence for Earth's rotation came in 1851 when French physicist Jean Foucault constructed a pendulum. A pendulum is a device that consists of a mass hung at the end of a string or wire that is able to swing freely. Foucault's pendulum consisted of a large iron sphere at the end of a long wire. Once in motion, the direction of a pendulum does not change. Yet the direction of Foucault's pendulum did seem to change. Since the pendulum could not change direction, Foucault concluded that it was Earth turning beneath his pendulum. If the pendulum were used at the North Pole, Earth would rotate in a full circle under the pendulum in 24 hours (15° per hour). In locations closer to the equator, the pendulum appears to take longer to move in a full circle. In most of the United States, it takes about a day and a half. You may have seen this kind of pendulum in a museum.

Another type of evidence for Earth's rotation came from the Coriolis effect. Recall that winds and water currents do not move in straight paths over Earth but instead move in curved patterns. Furthermore, the curve is toward the right in the Northern Hemisphere and toward the left in the Southern Hemisphere. The Coriolis effect can be explained only by the rotation of Earth.

Evidence for Earth's revolution comes from the stars. When viewing the stars over time, nearby stars appear to shift in position when compared with distant stars. Although you cannot measure this shift with your unaided eye, it can be measured

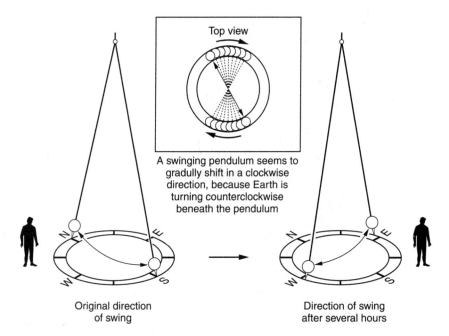

Figure 28-2. A Foucault pendulum is free to rotate as it swings. It will swing in the same plane as Earth turns beneath it. The period of apparent rotation depends on latitude.

with precise instruments. No such shift would occur if Earth did not orbit the sun.

Additional evidence comes from changes in the constellations. When we look at the night sky, we can identify star patterns. A group of stars that appears to form a pattern in the sky is called a **constellation**. There are 88 constellations that can be observed from Earth. Some of these constellations are Ursa Major and Ursa Minor (the Great Bear and the Little Bear, which include the Big Dipper and the Little Dipper respectively), Orion, and Cassiopeia.

The positions of the constellations appear to change throughout the year. Most constellations can be seen only at certain times of the year. The change in constellations is further evidence for Earth's revolution. We would see the same constellations all year long if Earth did not move around the sun.

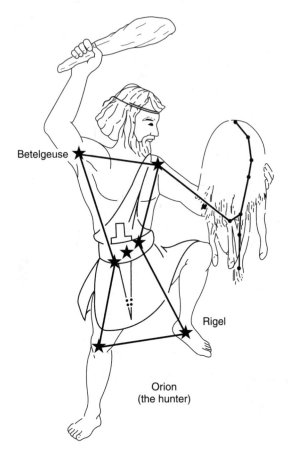

Betelgeuse

Rigel

Orion
(the hunter)

**Figure 28-3. Orion is the
brightest constellation
in the winter sky. The
bright stars, including
Betelgeuse and Rigel,
are easy to find, even in
urban areas.**

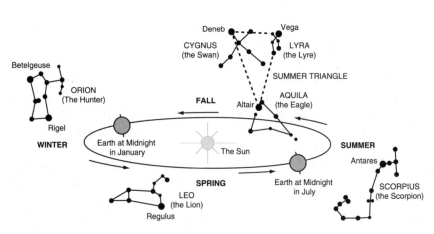

**Figure 28-4. As Earth revolves around the sun, Earth's nightside faces
toward different constellations. Only those constellations that face
Earth's nightside are visible.**

THE SEASONS

Earth's motion around the sun gives rise to the seasons. In Chapter 25, you learned about the seasons on Earth. Most other planets also have seasons. Mercury has none while Venus and Jupiter do not have obvious seasons. Now that you know a bit more about planetary motion, you can better understand why some planets have seasons and others do not. All the planets that have seasons are tilted on their axis. Seasons are caused by the tilt of a planet on its axis. Consider Earth as an example. Earth spins on its axis as it moves around the sun. Since the axis is tilted, at some points in its orbit one hemisphere is pointed toward the sun. At other points, that same hemisphere is pointed away from the sun. The hemisphere that is pointed toward the sun receives more direct rays of sunlight and has longer days than the other hemisphere. The combination of more direct sunlight and longer days results in the summer season. A hemisphere experiences summer when it is pointed toward the sun and winter when it is pointed away from the sun. Twice a year, neither pole is pointed toward the sun. These times are known as equinoxes, which means that day and night are equal all over the planet. (Look back to Figure 25-4 for an illustration.)

CHARACTERISTICS OF EARTH'S MOON

You have learned that several planets have moons. Earth's moon is of particular importance to us because it is the closest object to Earth in space. And it is the only extraterrestrial object that has been visited by astronauts.

The average distance to the moon is 384,403 kilometers. The diameter of the moon is 3476 kilometers. That is more than one-fourth the diameter of Earth. Gravity on the moon is about one-sixth the gravity on Earth. That means that on the moon you would weigh one-sixth of what you do on Earth. The moon has no atmosphere and therefore no weather.

It takes $27\frac{1}{3}$ days for the moon to revolve around Earth. The moon revolves around Earth in an orbit that is nearly circular. The point at which the moon is closest to Earth is called **perigee**. The point at which the moon is farthest is **apogee**.

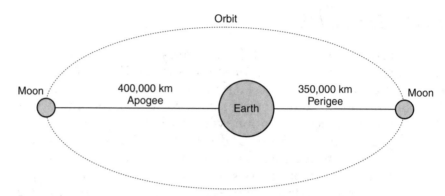

Figure 28-5. The moon's orbit around Earth. (Eccentricity exaggerated.)

The moon spins on its axis at the same rate that it revolves around Earth. In the 1960s, scientists got their first look at the far side of the moon via satellite. Since the moon rotates and revolves at the same rate, the same side of the moon always faces Earth.

The National Aeronautics and Space Administration (NASA) sent astronauts to the moon six times during the *Apollo* missions between 1969 and 1972. These astronauts left measurement devices on the moon and also brought back rock samples for analysis. One device the astronauts left was a small mirror. Scientists on Earth shined a laser beam at the mirror on the moon. The mirror reflected the light to Earth. Scientists then used the known speed of light to calculate the exact distance to the moon.

Another device they left on the moon was a seismometer, which detects moonquakes (like earthquakes). From the data recorded, scientists were able to draw conclusions about the composition of the moon. They determined that the moon is layered much like Earth. It has a crust, a mantle, and a solid core. However, they were unable to determine the exact composition of the core.

Features of the Moon

As seen through a telescope, the moon has dark and light areas. In the 1600s, Galileo thought that the dark areas were seas or basins filled with water. For this reason, he named them **maria**, the Latin word for "seas." Scientists now know that the lunar maria do not contain water. Instead, the maria are smooth plains with huge basins. Rock samples from maria are similar to the basalts in lava flows from the Hawaiian volcanoes. Long, deep cracks running through maria bedrock are called **rilles**.

The light areas are lunar **highlands**. These regions are mountainous and heavily cratered. The highlands appear lighter because their rocks, which contain much feldspar, are lighter in color. Most lunar craters were formed when meteoroids hit the moon's surface. Craters range in size from microscopic to hundreds of kilometers across.

The side of the moon facing Earth, the front of the moon, is made of about half highlands and half maria. The other side of the moon, the back, is mostly highlands and craters. There are not very many basaltic maria on the far side of the moon, which is never directly visible from the Earth.

The moon does not have a true soil. Instead, it has what scientists call regolith, which consists of loose rock materials. The lunar regolith is different from soil on Earth in that it does not contain water or organic material. And unlike soil on Earth, regolith was formed from the impact of meteoroids.

Lunar rocks are different from rocks on Earth in that they contain no water at all. They consist mostly of elements with high melting points, such as aluminum, titanium, and zirconium. They contain lesser amounts of elements with low melting points, such as sulfur and lead. And lunar rocks contain lesser amounts of elements that exist as gases, such as nitrogen and chlorine.

Phases of the Moon

When observed from Earth, the moon appears to go through a series of phases. There are two main reasons for the **phases of the moon**. One is that we see the moon as a result of sunlight reflected from it. The sun lights the half of the moon facing it. However, as the moon moves around Earth, an observer on Earth does not always see that fully lit side. At the new moon phase, for example, the lighted half of the moon faces away from Earth. So the moon appears dark from Earth. At the crescent phases, only a small part of the lighted half faces Earth. At the quarter phases, half of the moon facing Earth is lighted and half is not. At the gibbous phases, almost all of the lighted half faces Earth. A full moon occurs when the entire lit half faces Earth.

Figure 28-6. On this photograph of the moon's surface, darker areas are maria, and brighter areas are highlands.

Eclipses

As the moon moves, it can become involved in eclipses. An eclipse has to do with shadows. Both the moon and Earth cast shadows in space. A shadow is created when an object is placed in front of a light source, such as the sun. A shadow has two parts. The first part is called the **umbra**, which is the total shadow. The second part is the **penumbra**, which is a partial shadow surrounding the umbra. When the moon passes into Earth's umbra, a **lunar eclipse** occurs. A lunar eclipse can occur only during the full moon phase. Although full moons occur

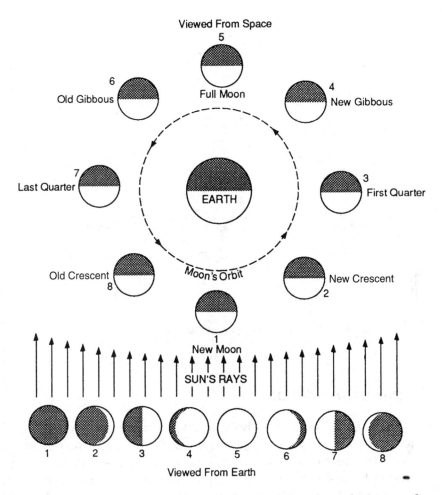

Figure 28-7. Notice how the lighted portion of the moon changes as the moon orbits Earth. This results in the phases of the moon.

every month, lunar eclipses do not occur at every full moon. The reason is that the moon's orbit is slightly inclined when compared to Earth's orbit. If the moon is above or below Earth's umbra, an eclipse does not occur. When the moon's umbra reaches Earth's surface, a **solar eclipse** occurs. This can happen only when the moon is at perigee. During a solar eclipse, the moon blocks all sunlight to a specific location, leaving the sky dark. A solar eclipse occurs only at the new moon phase.

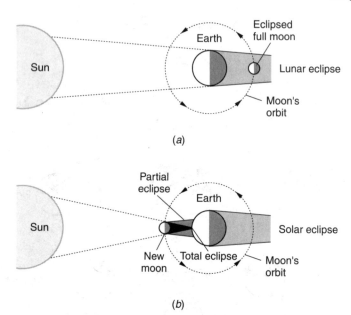

Figure 28-8. A lunar eclipse occurs when Earth casts its shadow on the moon. A solar eclipse occurs when the moon casts its shadow on a portion of Earth.

Origin of the Moon

No one knows for sure how and when the moon formed, but several theories have been proposed based on analysis of lunar rocks. The most widely supported theory suggests that the moon formed about 4.6 billion years ago when Earth collided with a large object. During the collision, some of the then liquid interior of Earth spilled out of the crust and entered space as a molten rock. The rock cooled and became Earth's moon.

LIVING ON PLANET EARTH

A native American proverb states that we do not inherit the Earth from our ancestors, we borrow it from our children. Of all of the legacies that we pass to our descendants, a favorable environment is one of the most important. But how can we best relate to our environment?

Many people believe that we have an obligation to protect our environment. The concept of stewardship places humans in a custodial role. It is currently estimated that humans move more earth materials in our construction projects than all the sediment transported by the rivers of the world. We also use farming practices that deplete soils and expose them to erosion. Both visible and invisible waste of the industrial economy can be found in the oceans. We may be changing world climates by venting waste gases into the atmosphere. We know that it is possible to reverse these trends, but do we have the social and economic will to do so? How alarming must the consequences of our consumer society become before we take effective remedial action?

We can cite many examples of new practices that help preserve the environment. Contour plowing (following level lines along the land) and selective placement of unplowed land have helped preserve soil. As a result of cleaner fuels and emission controls, the air in our cities is cleaner today than it was a generation ago. Upgraded waste disposal facilities and international pressure on shipping companies have reduced both freshwater and ocean pollution.

Other solutions elude us. Chemical wastes in waterways and oceans are often invisible to us, but some of them are very toxic, even in very low concentrations. Our increasing affluence and desire for inexpensive energy have increased the amount of carbon dioxide and other greenhouse gases in the atmosphere. The best ways to satisfy our growing energy needs by using alternate forms of energy are unclear. Solar energy is very expensive, and nuclear energy involves serious issues of public safety. Natural cycles are not understood well enough to enable us to predict future climatic changes without any human influence. Furthermore, it is likely that some regions would benefit from greenhouse-related climatic changes.

The Human Population

Population growth is a controversial issue. Scientists agree that the greater the human population, the greater the demands on our resources and our environment. Technology magnifies that environmental stress. As technology advances, individuals require and can afford more energy and possessions with each passing generation.

Before the Industrial Revolution, the combined ravages of complications in childbirth, communicable diseases (including epidemics), and malnutrition kept Earth's human population relatively low. Even now, in nations without elder care, parents need to have many children to ensure that some of them will survive long enough to take care of the parents in their old age. The United States, like other economically advanced nations, has medical advantages, food production, and social security that negate the need for large families. In addition, we lavish our children with expensive services, possessions, and educational opportunities. For these reasons the average family size in much of the Western world has decreased dramatically, and the populations of the wealthiest countries have largely stabilized.

However, the population of the less-developed nations, and therefore the world population, now six billion people, is still increasing. Can our planet support billions more? A generation ago, some scientists predicted that we would run out of food when the world population reached five billion. Not only are we now beyond that population mark, but the availability of food per person has actually increased. The use of fertilizers and biological engineering have revolutionized food production. This has led to greater per capita food production than ever before, in spite of an increasing world population.

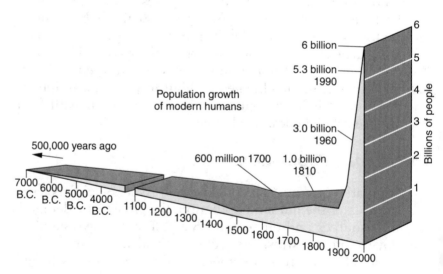

Figure 28-9. Since the Industrial Revolution, Earth's human population has grown very quickly.

Will the world population continue to increase at an accelerating rate of growth? This did occur into the 1960s, but the rate of growth has slowed since then, and the United Nations estimates that the world population will stabilize at 10–12 billion people in this century. If technology continues to develop, overpopulation is not likely to be a serious problem.

WHAT IS THE FATE OF THE PLANET?

In spite of the good news that history has brought us, perils still lie ahead. While economic forces are likely to limit population growth and provide for our needs, we should focus on the important goal of maintaining and enhancing a healthy and attractive environment. Economic forces alone are not enough. Creating parkland and cleaning litter from the environment have been initiatives largely related to our values and aesthetics.

We tend to see environmental quality in human terms. The first life-forms produced oxygen as a waste product. Yet, for the dominant life-forms today, oxygen is essential. Is oxygen good or bad? Was the change to an oxygen-rich atmosphere hostile to the earliest life-forms truly beneficial? Of course that depends upon your point of view. Clearly, as humans, we want to make the environment as favorable to us as we can. For the most part, that means preserving the natural environmental conditions under which we evolved. Another drastic change in Earth's atmosphere would probably lead to new conditions of stability, and it might be favorable to very different forms of life, but it would probably mean disaster for human civilization. Who is to say that change is bad? However, from the human perspective, the judgment is clearer.

WHAT IS THE FATE OF THE UNIVERSE?

If you throw a ball in to the air, it will rise to the top of its path and then fall back to the Earth. But if you could throw it fast enough to escape Earth's gravity, the ball would continue forever into outer space. (This account ignores the slowing effects of Earth's atmosphere.) Although it would slow down as it rose, any object propelled at a speed greater than the escape velocity would have enough speed to leave Earth and never return. The escape velocity for objects at Earth's surface is about 11 kilometers per second.

The universe also has an escape velocity. That speed depends on the diameter of the universe, now thought to be about 30 billion light-years, and the total mass of the universe, which is not known. Since we do not know the mass of the universe, we do not know the escape velocity. That is, we do not know whether the universe has enough mass to cause it to stop expanding and then begin to collapse at some time in the future, or whether expansion will continue indefinitely in spite of the slowing effect of gravity. On this critical value depends the fate of the universe. If the universe is massive enough, it will stop expanding and fall back into a superdense, hot mass. This would mean a fiery end to civilization, if our descendants still exist. But if the universe has too little mass to stop its expansion, spreading will continue indefinitely and our civilization will face a cold death as even the closest stars, like our sun, become too distant to support life on the planet. Alarming as both forecasts of the future may be, it would be billions of years before they occur. For the foreseeable future, the universe will provide the familiar environment in space that we have learned to value.

WHAT IS THE PURPOSE OF EARTH SCIENCE IN EDUCATION?

The final exam should not be your last contact with earth science. Throughout life we make decisions that are related to our planet and how we interact with it. As property owners we must decide where to live and how our properties are developed. As citizens we will vote on regulations about how private and public resources are preserved or exploited. We need to recognize and deal with environmental problems. How will we balance our desires for comfort and wealth with a need to preserve a favorable environment? We must understand how natural systems operate and interact. Above all, we must be conscious of how our activities will affect our environment and our future.

For these reasons, we also must keep ourselves informed about Earth-related issues as they develop throughout our lives. Knowledge is power. To maintain that power, we must maintain ourselves as lifelong learners by taking an active interest in issues related to our home planet.

QUESTIONS

Multiple Choice

1. How many hourly time zones are there on Earth? *a.* 15 *b.* 24 *c.* 60 *d.* 360

2. The Foucault pendulum provides evidence for Earth's *a.* seasons *b.* revolution *c.* rotation *d.* magnetism.

3. Lunar maria are *a.* huge basins *b.* vast seas *c.* tall mountains *d.* frozen glaciers.

4. During a lunar eclipse, the moon *a.* temporarily leaves its orbit *b.* moves in reverse *c.* moves into Earth's shadow *d.* casts a shadow on Earth.

5. Why do humans need to understand environmental systems? *a.* We are unable to change the environment unless we understand it. *b.* We need to preserve the beneficial aspects of our environment. *c.* If we understand the environment now, it will always be the same. *d.* It is likely that conditions on Earth are similar to those on other nearby planets.

6. What part of the environment are humans the *least* likely to change? *a.* the oceans *b.* the atmosphere *c.* soil and mineral resources *d.* deep interior.

Fill In

7. Earth is divided into _____ as a result of Earth's rotation.

8. A(an) _____ is a group of stars that form an organized pattern in the sky.

9. The moon is at _____ when it is closest to Earth.

10. Mountainous regions on the moon are known as _____.

11. Full moon, crescent moon, and quarter moon are _____ of the moon.

Free Response

12. Why is Earth divided into time zones?
13. What evidence can be used to show that Earth both rotates and revolves?
14. Why does the moon appear to have light and dark areas?
15. Explain why the moon cycles through various phases.
16. What is an eclipse?
17. How does the size of the population impact the planet?

Portfolio

 I. Create a poster showing how the motion of Earth is related to the seasons where you live.
 II. Keep a calendar for several months. Describe the moon by phase over time.
III. Make a list of five proposals to enhance environmental quality in your community or region.
IV. Construct a time line of both past and future events that have had or will have the greatest effect on our planet's human population.
 V. Find a person in your community who works to improve environmental quality. Describe that person's accomplishments. Suggest how you hope to apply your knowledge of earth science in the future.

Appendix
Table of Elements

Element	Symbol	Element	Symbol
Actinium	Ac	Erbium	Er
Aluminum	Al	Europium	Eu
Americium	Am	Fermium	Fm
Antimony	Sb	Fluorine	F
Argon	Ar	Francium	Fr
Arsenic	As	Gadolinium	Gd
Astatine	At	Gallium	Ga
Barium	Ba	Germanium	Ge
Berkelium	Bk	Gold	Au
Beryllium	Be	Hafnium	Hf
Bismuth	Bi	Hassium	Hs
Bohrium	Bh	Helium	He
Boron	B	Holmium	Ho
Bromine	Br	Hydrogen	H
Cadmium	Cd	Indium	In
Calcium	Ca	Iodine	I
Californium	Cf	Iridium	Ir
Carbon	C	Iron	Fe
Cerium	Ce	Krypton	Kr
Cesium	Cs	Lanthanum	La
Chlorine	Cl	Lawrencium	Lr
Chromium	Cr	Lead	Pb
Cobalt	Co	Lithium	Li
Copper	Cu	Lutetium	Lu
Curium	Cm	Magnesium	Mg
Dubnium	Db	Manganese	Mn
Dysprosium	Dy	Meitnerium	Mt
Einsteinium	E (Es)	Mendelevium	Md

Element	Symbol	Element	Symbol
Mercury	Hg	Samarium	Sm
Molybdenum	Mo	Scandium	Sc
Neodymium	Nd	Seaborgium	Sg
Neon	Ne	Selenium	Se
Neptunium	Np	Silicon	Si
Nickel	Ni	Silver	Ag
Niobium	Nb	Sodium	Na
Nitrogen	N	Strontium	Sr
Nobelium	No	Sulfur	S
Osmium	Os	Tantalum	Ta
Oxygen	O	Technetium	Tc
Palladium	Pd	Tellurium	Te
Phosphorus	P	Terbium	Tb
Platinum	Pt	Thallium	Tl
Plutonium	Pu	Thorium	Th
Polonium	Po	Thulium	Tm
Potassium	K	Tin	Sn
Praseodymium	Pr	Titanium	Ti
Promethium	Pm	Tungsten	W
Protactinium	Pa	(Wolfram)	
Radium	Ra	Uranium	U
Radon	Rn	Vanadium	V
Rhenium	Re	Xenon	Xe
Rhodium	Rh	Ytterbium	Yb
Rubidium	Rb	Yttrium	Y
Ruthenium	Ru	Zinc	Zn
Rutherfordium	Rf	Zirconium	Zr

INDEX